BE AN

InventHer

BE AN
InventHer
AN EVERYWOMAN'S GUIDE
TO CREATING THE NEXT BIG THING

MINA YOO & HILARY MEYERSON

SASQUATCH BOOKS
SEATTLE

Printed in the United States of America

SASQUATCH BOOKS with colophon is a registered trademark
of Penguin Random House LLC

23 22 21 20 19   9 8 7 6 5 4 3 2 1

Editor: Jennifer Worick
Production editor: Jill Saginario
Design: Bryce de Flamand
Cover and interior illustration: Andrea Floren

Library of Congress Cataloging-in-Publication Data is available

ISBN: 978-1-63217-275-4

Sasquatch Books
1904 Third Avenue, Suite 710
Seattle, WA 98101

SasquatchBooks.com

To my family—my husband, Mark, and kids, Kai and Mila—who didn't bat an eyelash when they learned that Mom had added writing a book to her already hectic life. Thank you for all your love, support, and encouragement in whatever I do (and for understanding that weekend naps are at the ground floor of Mom's hierarchy of needs). —**MY**

For Randy, my partner in inventing an extraordinary life. You told me once in a dorm room in Vermont that you were my biggest fan. It's always been mutual. —**HM**

Contents

Introduction:
Let's Channel Female Creativity

Women are the ultimate creators. That is the chewy caramel nougat heart of this book. But women are wildly underrepresented in the pantheon of product creators, and for no good reason. We believe that lots of women are walking around with genius ideas for inventions and products that we all need, but they don't make the leap to thinking about becoming an inventor—or an InventHer, as we're calling them. Why not? Historically, most inventors have been male (although there are a surprising number of women inventors who have given us some life-changing inventions), but we know much of that can be explained by traditional domestic roles, unequal societal power, and a long history of female disempowerment. However, women are taking the lead in some traditionally male-dominated careers—the majority of medical school and law school graduates are now women. We've got a ways to go toward proportional representation in many areas (hello, Congress, we're looking at you), and creating new products is just one. Just 8 percent of US patents filed each year list a woman as the primary inventor, according to the Institute for Women's Policy Research.[1] Luckily, this is one area where we as authors have some expertise, and we are committed to changing this. Women are making huge inroads in founding start-ups and small businesses, which have a more clear path of development. But inventing a product isn't something that gets a lot of attention, and few (men or women) have an idea of how to go about doing it. We know women bring different skills and strengths to the table than men in any business, and their contributions are invaluable. Everyone benefits from a multitude of perspectives and creative thought. We firmly believe that women are a huge untapped source of innovative product ideas—all they need is a road map to successful product development and creation. So we created one! And we think the fact that you picked up this book means you have the nagging feeling that you just might want to try to bring your invention into the world.

Why are women particularly suited to creating products? Because we are the ultimate problem solvers. Juggling everything from kids, work, household chores, and elderly parents (let's be truthful—many family and household tasks still fall disproportionately on women), we take action to make our and others' lives easier, coming up with ingenious solutions to life's problems, big and small. Sara Blakely was headed to a party when she realized she needed a better undergarment for a smooth look under pants—with a pair of scissors and some control-top pantyhose, the seeds of Spanx were planted. Joy Mangano was tired of getting her hands dirty mopping floors and came up with the Miracle Mop. The exterior fire escape was patented by Anna Connelly, who, we can safely assume, was looking for a way for people not to be burned alive. Marion Donovan invented the disposable diaper when . . . well, we can all figure out the inspiration there.

InventHers come from all walks of life and defy narrow definition by profession or demographic. Stephanie Kwolek was a chemist for DuPont when she invented Kevlar in 1965. Monopoly was invented by Elizabeth Magie in 1904. Magie was an untraditional woman who worked as a stenographer, secretary, writer, and comic before coming up with a game to educate people on her theory of economics and property ownership. Hedy Lamarr, the movie star of the 1940s, invented the technology that is the basis of today's Bluetooth, Wi-Fi, and GPS. The only difference between these inventors and you is that they took their ideas and ran with them. They brought their creations to life despite societal obstacles. Some had their ideas stolen and patented by others—this, ironically, happened to the Monopoly inventor. But it doesn't change the fact that each of them had an idea and created something. Lots of these women faced incredible challenges, but we don't think it should be so hard. Let's make it easier for the next generation of InventHers.

Women Are Crafty

Women create things on an extraordinary level, just not always in ways that make money. Think of all the crafters you know—those who knit or sew quilts or clothing or make jewelry, tote bags, pottery, or other crafty items. How about the chefs? Cooking is a form of creation, and the culinary arts have been an outlet for creativity for generations, when there were few others. Let's not forget about children—it's women who literally create those little tykes, and if that isn't enough, we still have to come up with science fair projects and unicorn-themed birthday parties with matching party favors. Moms are definitely the ultimate InventHers—we once watched a mom on an airplane create a toy for her cranky toddler out of two tiny water bottles, some coffee stirrers, and some tape—it worked like a charm. Women are natural problem solvers, which often translates to creating something. Is there a woman alive who hasn't created a makeshift maxi pad out of toilet paper and some clever folding?

We want to channel this creativity and point it firmly in the direction of commerce. Fun fact: women are responsible for 70 to 80 percent of consumer purchasing, both through direct purchases and through influence on purchasing.[2] Wouldn't it be awesome if women were the ones making and profiting from the products we buy anyway?

Why Us?

A few years ago Mina Yoo was a mom with an idea for a product: a clip—like a carabiner, but with a hook. She came up with the idea when, as a new mom, she decided to challenge herself to climb Mount Rainier. She got familiar with carabiners and thought, "These would be super useful in everyday life, but not quite as is. Something like this but with a hook would be like an extra set of hands for handling the gear of daily life." Just like that, the idea for the Heroclip was born.

Despite the fact that she had no design or manufacturing knowledge, Mina launched this first product, and several years of trials and errors as well as triumphs and learning ensued. But the end result was the creation of a design-oriented lifestyle-product company with goods successfully selling at large retailers like Ace Hardware, Bed, Bath & Beyond, and REI. Hilary Meyerson first came across Heroclip when reviewing outdoor gear for a magazine (she's also an expert in outdoor gear and the outdoor retail market) and was super impressed with it. She came in to help with the crowdfunding and marketing of the Heroclip, and they instantly bonded over the shared goal of women as successful entrepreneurs. Hilary wanted to tell the story of Mina's route to success, and Mina wanted Heroclip to grow even bigger.

As Mina's business expanded, she began to receive more and more phone calls and emails from friends, friends of friends, and more women from the extended network we all have. They all had the same question: "I have a product idea, but I don't know what to do. Can you tell me what you did?" Sheepish questions would always find their way into the conversation: "How much/how long will everything take?" Just like that, another idea was born: Hilary and Mina would share their knowledge and help other women follow the path to successful product creation. This book is the result.

The story of this book's origin and Hilary and Mina's friendship is another example of why women are uniquely positioned to be InventHers. Besides being creative problem solvers, we are the ultimate networkers. Women excel at developing relationships, and that is key to building a successful business. Long before LinkedIn, women forged friendships and relationships that kept their homes running, kids cared for, drains unclogged, taxes done, hair roots colored, and root canals performed. What do you do when you need to find a contractor who specializes in side sewers because your front yard is now an open septic system? You call the girlfriend network, because someone else has had this happen. Same if you need a referral for a doctor to treat your kid's ADHD, an after-hours veterinarian for your dog, or a roofer who will repair the damage from the last storm. In the professional realm, women look to other women for mentoring and advice. As a young attorney, when Hilary struggled with a brief, it was the women partners at the law firm who were her go-to resources for assistance. When Mina was in academia teaching and researching entrepreneurship, she also wished for female mentors—now she strives to mentor other young women. We believe that all women have huge female networks to tap into for resources and can use these connections to build a brand. Use this book and harness these networks for retail success and world domination. To paraphrase Beyoncé, "Who run da world? The ladies in your Facebook group."

For men, the cliché of business networking has been signing deals out on the golf course, or clinking whiskey glasses in a smoky cigar club. Hilary and Mina prefer some chilled pinot grigio and hashing out ideas over a kitchen table. And up until now, there wasn't a resource that reflected this more female approach. This is the book Mina wished she'd had when she launched her business. We want to help you come up with a killer idea, evaluate it against the current market, educate yourself on manufacturing, figure out how to pay for your new business expenses, and sell your product to the eager masses—all while keeping your sanity and sense of humor—because while you're learning to make a living at your business, you need to get on with the business of living. We don't want women to succeed in a man's world—we want to make it a woman's world. A woman's place is—well, wherever the hell she wants to be. Whether it's the home or the boardroom, there is no reason she can't be in both.

This book is meant as a starting point for InventHers, and not an exhaustive manual. Think of it as an overview of the entire process of creating the next great thing. Each chapter is about one step in the process—you could write textbooks about each individual step (and believe us, they are out there but are much less fun to read). We use real-life InventHers to illustrate how women have tackled each step, and we hope they will inspire you as they inspired us.

So pull up a chair and grab a glass—the wine (or your beverage of choice) is right over there. Let's get ready to talk products. The ice cream maker. The windshield wiper. The flat-bottomed paper lunch bag. The life raft. The dishwasher. The retractable dog leash. All invented by women just like you. The women's network is fully operational as we get you started on creating the next big thing—yours.

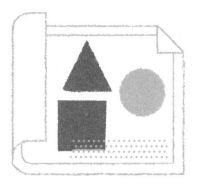
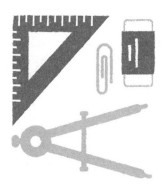

Chapter 1: The Initial Idea
The Baby Stages of Market Research

It might seem like every invention has an "aha!" moment story; that brilliant flash of insight, which then becomes a bestselling product throughout the world. When you are being interviewed at your local TV station about your wildly successful invention, you will want an inspiring story of how you stepped on Bubble Wrap one day and came up with a great idea for an intruder alert system based on popping small plastic spheres. But the truth is ideas that become great products don't always come from a single "aha!" moment. A product's origin story might not be so sound-bite perfect. Ideas can come from experiencing mundane daily problems, being a consumer unhappy with current choices, or just being a keen observer of the world around you. The success stories of other InventHers are not instances of idea gods randomly striking individuals with bolts of inspiration, acts that can't be replicated. Rather, anyone with an idea has the potential to become a success story if she works for it (and has this book by her side).

Like to Complain? You're Way Ahead of the Game

Inspiration is everywhere! For many InventHers, the initial baby lightning bolt of an idea comes from their own daily lives, when they continually face a nagging *problem* for which they cannot find a desirable *solution*. We say "desirable" because it is very possible that a solution exists, but for whatever reason it is not the right solution for you (and others like you). One of our InventHers, Leslie Pierson, noticed that her five-year-old son was coming home from day care with sheets and sheets of artwork. Her refrigerator door was covered in no time, and the artwork expanded to nearby walls. The problem? She didn't want to damage her walls (or the artwork) with tape or pins and wanted to be able to easily change out the art. The challenge was how to turn a wall into a magnet board for precious artwork without permanently changing it or incurring large costs. The idea for GoodHangups—a removable magnetic hanging system—was born to solve this problem. Excy InventHer Michele Mehl was a fitness buff (formerly a two-sport college athlete) who broke her leg and couldn't exercise at home the way she wanted. There were a zillion home fitness machines out there already, but they were large and expensive and not right for Michele. She wanted equipment small enough to be tucked in a closet but versatile enough to work out her entire body. The idea for Excy started percolating. The genesis of Mina's product was similar to Michele's and Leslie's. After having her first baby and then training to summit Mount Rainier, Mina realized that she was constantly lugging a lot of things, and often there was not a good place to set her stuff when she needed her hands to do something else (or just take a break). Although she found items like bungee cords, carabiners, and mommy hooks in the market already, she wanted an elegant, simple solution that could go from activity to activity with her, no matter where she was or what she was doing. What she really needed was an extra hand (at a price way lower than what an actual extra hand would cost), and the initial idea for Heroclip was born.

For your own first "aha!" moment or idea genesis, don't limit yourself to problems that only you face. Hey, you could have a completely problem-free life for all we know, but don't let that stop you from solving other people's problems! Although a great many inventions are inspired by our own lives, other inventions are driven by a desire to solve problems that affect society at large. Physicist Shirley Ann Jackson, the first black woman to earn a doctorate from the Massachusetts Institute of Technology, invented caller ID and call-waiting while working for AT&T Bell Laboratories in the 1970s. Nancie Weston, inventor of the Grayl water-purifier bottle, was inspired by the vast numbers of people in the world who do not have access to clean water. That led to a desire to provide them with an easy solution.

Done solving other people's problems through products? While our book focuses on physical products, inventions can be intangible products (like software) or not products at all. Sometimes an invention can come in the form of a *process*, or how something is done. Although Henry Ford wasn't the first to produce an automobile, he *was* the first to produce an automobile affordable to the general public. And he was able to do this by being the first to use a moving assembly line in which an unfinished product moved from one workstation to the next, getting parts added on until it became a finished product. Before the assembly line, the main model for making a product was a craftsman model, in which one person made the product from start to finish. Another case of a "process invention" is how things are sold. Amazon and online selling (e-commerce) is an example that affects pretty much anyone living on earth. Subscription boxes are another recent process invention. We also have the older example of direct selling (like Amway and Avon), where products are sold by trusted friends or at gatherings (Tupperware parties for our parents and Stella & Dot parties for us). Processes can also be patented—inventions can happen in any form!

Our point here is that problems are all around you, and you may very well have the solution, whatever the form it may take. The critical part of being an InventHer is to pay attention to the lightbulb moments, take them seriously enough to think about them beyond the initial "aha," and take the time to explore them to see if the ideas have legs. Check out the inspirations that led to our InventHers' creations in the table on page 5. We're not gonna lie, though: these were merely the first of the many "aha's" that eventually led to the materialization of this talented group's inventions.

The Problems Our InventHers Solved

Sarah Blankinship, Siva Patch	Needed a wellness/pain-relief transdermal delivery system for hemp/CBD that met exacting standards.
Chez Brungraber, Gobi Gear	Needed to organize cavernous backpacks so it would be easier to access gear quickly.
Amy Buckalter, Pulse	Needed a better-quality personal lube with an elegant delivery system.
Fran Dunaway, TomboyX	Needed high-quality, stylish clothing for an underrepresented demographic.
Judy Edwards, Squatty Potty	Needed to relieve constipation and discomfort related to pooping!
Lisa Fetterman, Nomiku	Needed a home sous vide device that was affordable and performed at professional chef quality.
Gloria Hwang, Thousand	Needed a stylish, lockable helmet recreational riders would like to wear.
Stephanie Lynn, Sweet Spot Skirts	Needed a universally sized sport skirt that could quickly snap on over athletic leggings.
Michele Mehl, Excy	Needed a portable, lightweight piece of equipment for exercise and rehabilitation.
Leslie Pierson, GoodHangups	Needed a wall-hanging system for artwork and other 2D items that wouldn't damage walls.
Sirena Rolfe, Tempus Hood	Needed an alternative to umbrellas and jackets to protect hair from rain.
Nancie Weston, Grayl	Needed a simpler way to purify water in the outdoors.
Mina Yoo, Heroclip	Needed an "extra pair of hands" to tote all the gear that comes with daily life.

Get Started By Taking This Easy Step (Seriously, It's Easy)

After your first "aha" moment, you may find yourself thinking about your idea all the time (or not at all, which is telling in and of itself). Just like you can't stop thinking about that truly special first date, when you have an idea that is truly compelling, it will stick with you (in a good way). Even when you tell yourself that you have too much going on in your life, that you know nothing about developing / making / marketing / selling a product, that surely someone else must have thought of it before—even when you have gotten distracted by life and forgotten about it for a few days or weeks or months— the idea will come back to you. Then you'll hear a story about an InventHer who used to be a fax saleswoman who came up with a body shaper that sold so well she became one of the richest women in the United States, and a tiny voice will say, "Why not me?"

Yes! Why not you? The mere belief in this possibility separates you from the majority of people who have a "great idea" and places you miles ahead of them in the journey of inventing. Let's hold on to this feeling of possibility and start taking your idea seriously. Seriously enough to start spending the time to really explore and assess its merit. In the spirit of committing to this (potentially) brilliant idea of yours, it's time to take an important first step . . .

Write It Down

And no, you don't have to get a special notebook for this step. Find whatever piece of paper you have handy and jot down a sentence or two describing what the product is/does and a sentence or two describing the problem it solves. You can use an electronic device too—we love voice recording ideas in the car—but paper is still our favorite "first draft." For good measure, draw a quick sketch of your future product. This is not some voodoo magic stuff, and no fairy godmother will be doing a "bibbidi-bobbidi-boo" to turn your idea into a bestselling product. However, there is power in writing out things you want to achieve because it *commits* you (in fact, we are going to go write down "Get at least twenty minutes of exercise every day" on a piece of paper right now).

> *Example*
> **Product:** A three-in-one travel pillow with a hood attachment to cover your eyes and keep your head warm while keeping your neck supported when you are upright and dozing.
>
> **Problem:** People don't want to carry around a sleep mask, a hat, and a pillow on their travels.

That wasn't too hard, right? We'll get more into marketing and market research in future chapters.

 HOT TIP: There is no substitute for paper for recording your first idea or draft.

Find Out What's Out There

Now that we are taking your idea seriously, let's spend a bit of time doing some comparison shopping. Use your friends Google and Amazon to find products on the market that might be able to do what your product does. Nothing will be as awesome as your product, of course, but what *could* someone use instead of your product to address whatever problem your product will solve? This is where research comes in.

A different way to think about other products on the market is, "What differentiates you from your competitors?" If you can't answer this question in two sentences or less, go rethink your product. It should be crystal clear. The *why* of your product is your *value proposition*. It's what you're bringing to the table. If you haven't built a better or at least a different mousetrap, you're not adding value to anyone. Go back to the drawing board and figure out what will make your customers part with their money. This "why" question will also be critical when it's time to market your product—you're already on your way to developing a marketing campaign and messaging when you are clear on what your differentiator is.

After you have spent a few hours with Google and Amazon, chances are that you will discover a product kinda/sorta like yours somewhere in the world. Don't be discouraged! Using the travel pillow example, you can quickly see it's a crowded field. But that doesn't mean there is no room in the world for *your* pillow. The important thing is that you figure out how you are going to *differentiate* your product from all the other travel pillows out there. There are many choices of travel pillows—why should anyone buy yours? How is it better? Is it cheaper? More comfortable? Made from sustainable materials? More convenient to stow away? More beautiful? Is there any reason that existing products do not fill a need that your pillow does? At the end of your Googling, you might conclude that there are too many similar products and your product is not different enough to be desirable—but better to know now!

Here's a little cheat sheet of things to think about when you are researching existing products and thinking of how your product can be different.

Features: What are some new/unique features your product will have over the existing products? They can be additions (a hood on a travel pillow), subtractions (a phone with no raised keys, or a bicycle with no handles), or some combination that has never been seen together before (an inflatable solar-powered lantern, or a French press mechanism for a water filter). Or they can be improvements in efficiency or effectiveness. Does your product do something others can do but faster? With less work? Or take up less space? Easier to carry or move?

Price: Price can be a huge differentiator. If your invention is really about an improvement in process so that you can offer the same product as everyone else but at half the price, trust us, your product is going to rock it. Be careful, though, if your idea is to compete on price. Unless there is some compelling reason that you are able to sell your product for less (like some new manufacturing or marketing process that you developed), differentiating your product based on a lower price is a strategy that is not only unsustainable, but also can lead to negative perceptions of the quality of your product.

Material: One way of differentiating your product is by using a material that hasn't been used in a product like yours before. Hilary, being a writer for multiple outdoor magazines, is constantly sent "new" and "innovative" products that, frankly, look like everything she has already seen before. But upon investigating a little closer, she often finds that the "invention" in a particular product is in the material used, such as the silver antimicrobial fabric that makes your workout clothes basically odor-free (cheers for this invention).

Design/color/pattern: Design, colors, and patterns may seem like superficial elements of a product, but product companies can be built on these factors. There is a reason that copyrights for creative work exist—design can be a critical component of the novelty of a product. (Did you know that patterns like "camouflage" can be copyrighted and/or trademarked?) Companies become known for particular designs or color schemes, giving them a leg up in differentiating themselves from makers of similar products.

Social responsibility: Your product's, or more accurately your company's, commitment to social responsibility may also be a differentiating factor, especially if a feature of your product gives back to society in some way. Glassybaby, a line of handblown votives started by a Seattle InventHer, gives 20 percent of its sales to cancer research. FinalStraw, a reusable, retractable metal straw that presold nearly $2 million worth of product in 2018, is all about reducing landfill waste created by disposable straws.[3] Other companies use recycled water bottles or sailcloth or fishing nets as materials for their items—they do good while differentiating their products.

Quality: We talked about how price can be a differentiator, but quality (which often accompanies a higher price) can be as well. Sell a premium product with exceptional quality and you will stand out in a sea of products with similar functions.

There is a reason that people buy Apple iPhones in droves when there is an abundance of other smartphones in the market and why people purchase high-priced insulated metal water bottles over cheaper plastic ones that presumably also hold water just fine.

Previously untapped market: Another differentiating factor can be to whom you sell. Fran Dunaway of TomboyX developed underwear, just like many other companies, but made her mark by designing it to appeal to gender-neutral, body-positive buyers who were not looking for traditionally "female" undergarments, a demographic that was not being addressed by traditional underwear companies.

Production method: Another form of differentiation can be the production method you use to make your products. We have all seen and bought jewelry because it was handmade. We may have bought certain products over similar ones because they were "Made in the USA" or made locally by neighborhood moms or because they were made in an environmentally safe facility. Increasingly, buyers are interested in how things are made, over and above the functional values of the products themselves, and a socially appealing method of production can be just the thing that differentiates you.

Packaging: This may seem superficial and trite, but if you think your product is going to be at brick-and-mortar stores, a novel way of packaging can differentiate your product from similar ones. A case in point is the bottled water industry. We don't know about you, but we really can't tell the difference in the product itself between the various water brands. How we can tell two brands of bottled water apart is through the packaging—the shape of the bottle, the cap, the design of the label, and the colors (oh, and we noticed that pricing is a differentiator for water as well—five-dollar bottled water, anyone?). While we are still believers of not judging a book by its cover, and while we do not recommend packaging as the only differentiating characteristic of your product, we do know that packaging can at times be the single most important factor that separates the product that's flying off the shelves and the product that is gathering dust on a lonely peg.

Affiliation: Sometimes who you associate with can say volumes about you, rightly or not. The same is true with products. Some products' only significant differentiator is with whom or what they are affiliated. Nike's Air Jordan stands out among all sneakers by virtue of being endorsed by one of the greatest basketball players in history. The "George Foreman Grill" was not invented by George Foreman but by an inventor who was able to capitalize on George Foreman's name and face recognition through this affiliation. Another supreme example is, well, Supreme, a brand that started as skater apparel and now attracts hipsters of all kinds. Products developed in collaboration with Supreme are quickly differentiated from similar products and frequently become highly desirable collectibles. From a consumer perspective, there is, again, a difference in price—a Supreme knife costs $105, while the exact same knife without the Supreme affiliation is $35.

It is critical that you spend a good amount of effort checking out your competition—the products that may be perceived as having similar features and benefits as yours—and figuring out whether your product can be differentiated in one or more of the ways we've described. If you find that you do indeed have a unique product, you can move forward with more confidence. If you discover that your idea probably is not different enough to cut through today's crowded marketplace, it may be best that you move on to other ideas (and we're sure you have many!).

Let the People Speak

You've already done research on your competition, and now it is time to talk to some people to see what they think of your idea. Actually, talking to people is so important in developing, marketing, and selling a product that we think it is always time to talk to people about your product and get honest feedback. So, it may be more accurate to say it is time to talk to the first five of all the people you will talk to in the life of your product.

If feedback is so important, why just five? You are, of course, welcome to talk to more if you have the opportunity, but we think that at this early stage, when your product concept is merely a sketch and a couple of sentences, five will suffice in giving you some sense of how much time you want to invest in this journey. We'll call these your First Five. Of all the people you will talk to, these five may be the most difficult but only because they are the first. They will be the scariest because, well, what if no one else thinks your brilliant idea is brilliant? What if they ask questions to which you have no answers? (This is very likely since you just started this journey!)

What if your research wasn't adequate and it turns out your product actually does exist somewhere? Your conversation may also be awkward if your research doesn't reveal any direct competition to your product—there is a real possibility that your product doesn't exist yet because the world doesn't need or want it right now (sounds harsh, we know), and you will have to watch as the person you are speaking to is racking her brain to figure out a kind way to say this. The hardest part is going to be getting your feedback giver to take your product concept seriously enough to give you thoughtful and useful feedback. This is nothing personal; just like you might not have initially taken your idea seriously because being a "product inventor" seems so out there and unreachable, your friend may not yet realize that you are dead serious about the potential of your product. Make the most of your conversations with your First Five by using the guide on page 16.

Besides bracing yourself for all kinds of feedback—good, bad, and neutral—who you pick for your First Five is also critical. You should do your best to pick people to whom you feel comfortable divulging your brilliant idea without fear that they will steal it and make it their own. At this stage, thoughtfulness and honesty are more important than the feedback givers being actual potential customers of your product. However, they should at least be able to identify with the kind of person who will buy your product (for example, a dad might not be the actual consumer of your new breast-milk pump, but he may be familiar with how it works, the challenges, and so on).

Tough critics will make a better product. We can't emphasize this enough. Find your friends and family members who are known for being brutally honest and ask them what they think. If you have teenage kids, they are helpful here—they are skeptical about everything and not prone to saying things just to make you feel good. If they give you anything better than "It's OK, I guess," you know you have a winner.

Even with all this feedback, you may still not get the full answer to whether your product will actually sell (and sell enough to make it worthwhile). Asking someone whether they *would buy* a product is different from having them actually *pay* for it. You won't know until you have made that jump and brought your product to market, but all good things take a little leap of faith, right? Crowdfunding is an effective way to see whether people want your product enough to pay for it before you actually go into mass production. With a crowdfunding site like Indiegogo or Kickstarter, people will prebuy your product based on your prototype. We'll cover crowdfunding more thoroughly later in the book.

The Quick and Dirty Patent Search

While you are doing research on competition, especially if you don't find anything remotely similar to your product, you might wonder whether a product like yours has been thought of but is simply not being sold (yet). To see if anyone has already patented something similar, you can do a quick online patent search. Again, Google is your friend. Go to Patents. Google.com and use the search tool and keywords relevant to your product. Easy! We do have to put the disclaimer here that we are not patent attorneys and no online tool (not even Google) can replace the work of a flesh-and-blood lawyer. But, hey, if you find 10,000 patents for products similar to yours, it might give you another data point to consider in your decision to pursue your product idea.

Preparing for Your Conversation with the First Five

WHAT TO BRING:

- your sketch, description, and list of who will use the product

- five products you found through your research that are most similar to yours, and notes on how your product is different

- a notebook or tape recorder to capture comments (see page 246 for a worksheet you can photocopy)

- an open mind

QUESTIONS TO ASK:

- Have you seen a product like this before?

- Do you think it would be useful for a _____? Why or why not?
  <small>WHO YOU THINK WILL USE THE PRODUCT</small>

- Would you pay for it?

- What would you pay for it? (This question will be more useful later on, but since you have the ear of five people, why not ask it?)

HOW TO ASK FOR FEEDBACK

Start with something to the effect of:

"Hey, I was doing _____,
WHAT INSPIRED THE IDEA

and I had this idea for a product that does _____,
WHAT IT DOES

which will help do _____.
HOW IT WILL MAKE WHATEVER YOU WERE DOING EASIER

I spent some time researching it and seeing if there is anything like it already, but I think my product is different enough to look into it more. Will you give me some feedback? I'm at a super early stage and just beginning to think about this, so don't worry about hurting my feelings, and feel free to give me your honest thoughts."

A BIRD'S-EYE VIEW OF THE BIGGER PICTURE

If, after your conversations with the First Five, you are still excited about your product, the next thing to do is look at demographic and consumer behavior trends. We know that this sounds academic, but having a good idea of where the buying population is headed in terms of size and habits will help you decide whether your product idea is worth pursuing. The Bureau of Labor Statistics is a good source for finding out how many people have what kind of jobs and what levels of income. The US Census tells you about sizes of households and trends in aging. If your product solves problems for those making over $1 million per year who are older than sixty-five years of age, don't you want to know the size of your total customer pool?

Depending on your product, it is also likely that there are other resources for industry data. For example, each year the Outdoor Industry Association publishes various data—how many people get outdoors, what kind of gear is selling for how much, and trends in sales. Turn again to your friend Google to find organizations or associations that collect data in the industry your product will enter. Yes, we know that looking at data and statistics is not as exciting as improving your product sketches or even—*gasp*—starting to work on a prototype. But a few simple visits to data-rich websites can help you determine whether you want to pursue your idea. See our Resources section on page 233 for some useful data sources.

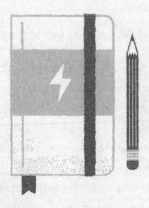

A Final Word

This chapter was all about figuring out whether the product your genius mind has conjured up is one that the world needs and wants. The homework we gave you will not guarantee that your product will be welcomed with open arms. What it *will* do is quickly give you a sense of not only your competition and whether at least five people think your product idea is a good one, but also, perhaps more importantly, how you feel about your own idea. The effort and general tenacity required to take the genius idea from your head and put it into the hands of your customer are not to be underestimated. In this book there is much more homework to be done. If the second person you speak with is lukewarm on your idea and you find yourself saying, "Whatever, I'm too busy to develop a product anyway," that speaks volumes about how much passion and conviction you have in the product idea. If you find yourself saying, "Hmm, I wonder how I can tweak my idea to be more appealing and useful," or "Hmm, what can I do to better explain the benefits of my product?" then it is a different story.

Doing early research—whether with Google or with your friends—is as much about you as it is about the product. We are big fans of learning quickly and either forging ahead or moving on to other endeavors. We are also big fans of doing what we feel passionate about. We all know that we have five hundred things we could be doing at any given moment, so why waste time on something you don't feel compelled by, right?

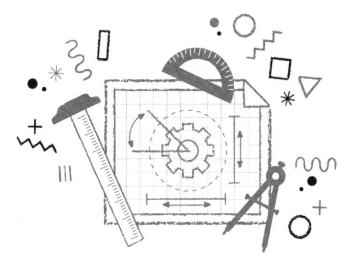

Chapter 2: Prototyping
Let's Get Crafty

While honing, modifying, pivoting, polishing, pulling apart, then putting back together the product idea you can't stop thinking about, a prototype is tremendously valuable. Let's face it, you could be an extroverted wordsmith and the former star of your high school debate team, but when you need feedback on your idea, there's no replacement for a prototype.

Designing and prototyping can be a challenge for "idea people"—it's get-your-hands-dirty-pretend-you're-crafty-and-make-an-attempt-at-building-it time. Some big thinkers shudder at the thought of a glue gun, a welding torch, or a sewing machine. But prototyping is crucial to product development. While very few people take even the most basic step in prototype making, the cocktail-napkin sketch is simply not enough to get a realistic idea of how the product will work, let alone production challenges and costs.

Don't be scared of prototyping—you've been doing this since you were a kid!

It's true that to really gauge whether your product idea has potential, you *do* need to have a fairly realistic prototype—and that requires some specialized skills (for most things). However, we suggest that you start simple and do whatever you can on your own using readily available materials before plunking down big money to get the specialists. The prototyping process can start out as rudimentarily as cutting up paper and gluing the pieces together, or sewing pieces of fabric together, or taking apart existing products and using the parts to create something totally different. Taking even this first step will put you way ahead of where the vast majority of people with ideas end up: while shopping at Target one day, they spot "their" product on the shelves and ask themselves why they didn't do anything about it years ago. Don't let this happen to you!

So what exactly is a prototype? It's your idea brought to life in three dimensions. While a picture is worth a thousand words, when it comes to product invention, a prototype is worth a million. It's what you'll show critical people like future customers, investors, manufacturers, lawyers, marketers, and others you will need when it's time to produce on a large scale. Making the prototype will also allow you to test, tweak, refine, or reimagine your idea (or ditch it and move on to another of the many ideas fighting for attention in your head). A prototype moves theory to reality, and it often looks quite different when it gets there. Also, as an InventHer, there is nothing quite like the thrill of actually holding what you dreamed up in your hot little hands.

Just so you're not cowed by the idea of creating a physical prototype, we're going to walk you through Mina's prototyping process for what would become the Heroclip.

CARDBOARD PROTOTYPE

Mina built her first prototype out of cardboard. She had her rough sketch and, armed with an X-Acto knife, she made the first 3D incarnation of Heroclip. After a couple of false starts and a couple rounds of cutting out shapes, she could already begin to see her product come to life. She attached two pieces of cardboard using a metal clasp she had cut out of an office envelope. Holding it in real life gave Mina the first impression of sizing. She could hold it up to various items and see how it would hang on things like tables, tree branches, strollers, and bar counters. It now became clear how large the product needed to be to function as she wanted. Mina adjusted the size a bit and soon had her rough draft for proportions.

INSTAMORPH PROTOTYPE

Unfortunately, this cardboard prototype also showed Mina that the design would not provide enough stability for the kinds of uses she had in mind. So, she modified the design a bit and decided it was time to step up her game and upgrade from cardboard to plastic. After some Googling (an InventHer's best friend), Mina invested seventeen dollars in a product called InstaMorph. Basically, you drop little plastic beads in hot water, they become moldable, and then you shape them into any form you want. Online, she saw all kinds of beautiful things that were made out of this product and was super excited. Well, her parts didn't come out as smoothly as all those online, but with a bit of superstrong glue and a tiny door hinge from Home Depot, it was enough for her to realize that her design had potential. The functionality seemed to be OK, but she couldn't hang any weight off it to replicate real-life situations due to the relative flimsiness of the material.

Nondisclosure Agreements

What is your most valuable asset in the development of a new product? Your 3D printer? The raw materials? The espresso machine in your office? Nope. It's your *idea*. It's intellectual property and it's valuable. The last thing you want is someone stealing it while you search for skilled team members to help you get your prototype off the ground. Who is to say that the great metal fabricator you hired isn't going to turn around and market it himself to XYZ Corporation? You need to have anyone involved in the development process sign a nondisclosure agreement (NDA), ensuring their confidentiality and waiving any right to the product due to their contributions. Also, make sure you have a good consulting agreement in place that states that all work performed by your designer (or service provider) is yours. A product designer you hire in the late stages may claim her tweaks give her a right to a portion of the proceeds. Get this wrapped up in advance—anyone you hire is a contractor or employee, not a cocreator. And make sure the agreement has real teeth for any violations. It's best to hire a lawyer for this, but at the very least get an NDA off the internet and have all parties sign it before they get down to work.

IRON PROTOTYPE

For the next iteration, Mina turned to a guy named Tabasco. She happened to see a picture of a friend posing with an ornamental metal mustache and asked him who had made it. Luckily, the blacksmith—Tabasco—was local. Mina knew she needed to eventually move to a metal prototype, and this was something she couldn't do in her kitchen. Tabasco (in his former life, he had played in a band with two lead singers named Ketchup and Mustard, and the nickname stuck) knew how to bend metal to his will and, armed with Mina's little InstaMorph model, he bent some iron into the basic shapes and curves she needed for her vision. The metal looked pretty rough, but with a little spray paint and a door hinge from the previous prototype, Mina could see a crude version of Heroclip come to life. She was ready for the next step.

GOING PRO

Here's where Mina hit the first major prototype obstacle. Buoyed by her progressive successes, she was ready to go pro. She went to her trusty research assistant Google, found three product-design firms in Seattle, and set up meetings with all of them. Sadly, one of them didn't even email her back. Another, after the first meeting, wrote to say that they were not sure they had time to work with her. A third sent a proposal for a total of . . . $50,000. *What????* After she recovered from the sticker shock, Mina read the remainder of the quote, which sent her into another tizzy. The price included just two revisions, and *there was no guarantee that the design could be manufactured.* Sure, a design firm would give her access to an entire group of talented individuals, but at this point all she had was her painted metal hook, and she wasn't even sure whether the product was feasible. It seemed like the product-design firm wouldn't be of help in figuring that out. It was disheartening to say the least and enough to drive Mina to the nearest bar.

 HOT TIP: Be confident! One reason the product designer Mina reached out to might not have responded was Mina's wishy-washy email saying she was "thinking of" pursuing the idea. If you want to be taken seriously, confidence matters.

FINDING A PROFESSIONAL PRODUCT DESIGNER

A *lot* of people do not want to work with inventors who have only a vague idea of what they want and must stay within a limited budget. Before taking up your time (and theirs), ask if they have worked with first-time inventors. Also, make sure they parse out each phase of design as narrowly as possible so that you will know quickly if they or your product idea is not working out. For example, ask a firm to commit to ten hours initially to come up with some sketches of concepts. If you like what they do, you can continue with them. The last thing you want is to receive a huge bill at the end of a six-month design period and end up wishing you had known what they had worked on so you could have put the kibosh on it sooner.

After the first sticker shock, Mina decided to forget the idea of getting a design firm and to forge ahead with a freelancer. This time, she started out on Craigslist and a site called Coroflot, which specifically features designers looking for jobs and gigs—there were a lot! The price wasn't right, though, and Mina was just about to reach out to design students at a local art school when she found an affordable product designer for thirty dollars an hour through a friend (put it out there to the universe—sometimes, if you mention your specific needs to a friend, she might know just the right person). The designer didn't have a ton of experience, but we believe in giving people a chance, especially at the right price. Sadly, although she helped Mina come up with some alternate designs, in the end she gave up (in the nicest possible way) and told Mina that she lacked the mechanical-engineering background to provide the type of solid, stable product Mina wanted. Who knew that Mina's little product would need a mechanical-engineering specialist?

At this point, Mina contemplated throwing in the towel. She had a ton of other things going on already and wondered, "Do I really need to do this?" She continued hiking and going out with her kid, and she could not stop feeling that somebody had to make this product. This handy little gadget was on her brain with the question "Why not me?" Thus began the hunt for an experienced designer with a mechanical-engineering background.

3D-PRINTED PROTOTYPE

Enter Joe, a freelance product designer who had cut his teeth designing snowboards and had his own line of luxury dog beds. One of his references had told Mina that Joe was "wicked smart," and sure enough, he was. He could look at various materials and the design and determine how much weight Mina's future product could hold, what the stress points were, and where the optimal pivot point for the clip would be located. Joe charged Mina ninety dollars an hour and spent approximately thirty hours getting the product to a point they were both comfortable with. His detailed drawings and specs allowed Mina to get a 3D print of the model (if you've never seen a 3D printer work, you are missing out on one of the coolest appliances around—it *prints* in plastic!). Mina's happy-dancing feet reverberated throughout the neighborhood when she saw that the prototype worked exactly as hoped and could hang anywhere.

After a couple more months of tinkering and improving (resulting in several more 3D prints), and a couple of informal focus groups to see what people thought of the product, Joe and Mina were ready to get factory-produced prototypes on which they could test weight loads and get quotes for production. To help them decide on the material for the end product, they ordered prototypes in aluminum, zinc, and plastic. So that they could perform weight-loading tests, they ordered three of each (as you'll see later, you are going to break some prototypes in the process). Aluminum ended up winning after they weighed the pros and cons of each material. Zinc was harder and more scratch-proof (zinc is what makes up the shiny hardware on your purse), but it was heavier and ironically gave the impression of looking "cheap." Plastic was cheaper to make but looked weak and actually couldn't hold as much as the metals.

Summary of Costs

DESCRIPTION	COST
DIY prototypes	Less than $50
Prototypes in 3D print	Less than $500
Factory prototypes	Less than $5,000
Manufacture-ready prototypes	Less than $50,000

3D Printers

Just a few years ago, these seemed like something out of a sci-fi movie. Now they're something you can pick up at Costco. Just like the arc of most technology, costs are coming down as capabilities are going up. The best-case scenario is to encourage your designer to get one, if they don't have one already (Mina's designer invested in a good one, and he always sends her a video of her product being printed; it's like seeing a baby being born, but much cleaner). Whatever you buy will surely be outdated in two years, so we wouldn't advise making an investment equivalent to what you'd pay for a car. Also note that a 3D printer can only print contiguous areas. So for Heroclip, Mina's team actually printed out three separate parts and then glued and screwed them together.

Another option for 3D printing is to email your design to a makerspace, have them print it for tens of dollars (depending on the size of your product), and pick it up when it is ready. Before her team got their own 3D printer, Mina used a makerspace as well. She wasn't initially worried about an NDA because her product was in pieces and it wasn't immediately evident what the finished item would look like. Plus, we think that the folks who work at makerspaces have so many of their own designs and enough respect for creativity and design that they wouldn't go stealing ideas, but that's your call to make. More cautious people will have different parts of their prototype printed by different places so that no one will get the complete picture. To see if your town has a makerspace, Google "3D printing services" or "makerspace."

The Regulatory Landscape

Some products are going to need a prototype for an additional reason: regulations. If you're making a car seat for kids, you're gonna need some approvals. The Consumer Product Safety Commission has a website devoted to consumer safety regulations, which has specific guides for different types of products[4]. (Child and infant products are some of the most tightly regulated out there, with good reason. For example, a full-sized crib has twelve separate requirements it has to meet just for the physical assembly of the product, and then a set of requirements for paint, registration, labeling, and the chemicals used to treat the surface.) These sorts of tests obviously demand professional help (see Resources, page 233).

All in all, in nine months, Mina spent in the low tens of thousands of dollars for the various prototypes and various professional help. Yes, this is a big investment, but all the stuff Mina got for her investment was a heck of a better deal than $50,000 for a pretty picture. By the time Mina had spent this much, she had a *manufacturable* design, prototypes to show buyers (in fact, she took the factory prototype to a trade show), and the ability to perform load-bearing tests on three different materials. Mina's experiences and the table on page 28 should also tell you that it is much, much smarter to do all your experimenting, tweaking, and modifying earlier in the prototype stage; it only gets more expensive to make changes as you get closer to a factory model.

Your Product Is a Snowflake
(But You Already Knew That)

Mina's route to her prototype might not be identical to yours. Your mileage and curves in the road, so to speak, will vary. But don't be discouraged along the way, and don't feel that because you can't afford a $50,000 design firm you won't have a solid prototype. If you're not happy with what you've made, keep reaching out to people with the skills to get it further down the road. Everyone has a Tabasco in their network; they just don't know it. Keep talking about your product, since you never know who might know whom, and look on Craigslist and Google to find people you need. Make sure that you email them with as much detail as possible so that you can see whether they are a good fit before you spend your time meeting with them. You don't need to tell them what the product is, but do let them know exactly what you need. Mina, for example, might have said:

> Hi, I am a new inventor, and I am working on a metal product. It is about the size of a climbing carabiner but requires some additional hinges. The material doesn't have to be as strong as a climbing carabiner because it won't be used to hold humans. The product might also need a rubber part as one of the components. I also need the product to have a beautiful finish because it will be used not just as a gadget but as an accessory. It has to be light but strong. Please let me know if you have worked with these kinds of materials and conditions and whether you have worked with individual inventors. If so, I would love for you to sign an NDA (nondisclosure agreement) so I can tell you more about it and learn whether it makes sense for us to work together.

The amount of time and financial resources it will take to make a solid prototype will vary. In the grand scheme of things, Heroclip is a relatively simple product. Your product may be a lot more complicated, involving not just mechanical engineering but also electrical and software engineering. The Nomiku and the Pulse, creations by two InventHers in this book, are examples of more complicated products. And complicated products take more time and money (shocking news, we know). Similarly, the level of innovation will vary across different product ideas—some inventions will be major innovations, and others will essentially be evolutions of existing products—and this will also affect time and costs for initial product development, as well as iterations and testing.

Prototyping 101

Here's a summary of steps to help you get to the prototype of your dreams.

- **Start with a drawing.** There's a reason the cocktail napkin has been the birthplace of many great ideas. The act of sketching helps you think through the product. Don't be intimidated by your lack of artistic skills. What's important is that you are thinking as you sketch. Problems will immediately become apparent. Paper is cheap, so use lots until you have a good rough idea. If you are computer savvy, build a virtual prototype using computer-aided design (CAD) tools, but don't be intimidated by this if it's beyond your skill level. You can always find someone to do this later. And lots of really slick CAD designs never make it to production.

- **Build it somehow.** Cardboard, InstaMorph, coat hangers, fabric remnants, broken vacuum cleaners, Play-Doh, wood, plastic—use whatever it takes to make a rough basic prototype so you can show your vision to others who can help you get it to the final prototype. Don't be afraid to borrow parts from different products.

- **Get a better prototype made.** If you are making a physical product that is not textile based, a 3D printer is your new best friend that you will count on over and over again during the development process. We cannot overemphasize how important this step is. You will save gobs of money by realizing early on whether your design is workable, and your final 3D-printed prototype will be good enough to provide the proof of concept you need to have some assurance that the product can work as intended. It will also help you optimize your production and uncover inefficiencies that weren't obvious at earlier stages. For electronics or more complex devices, you're going to have to find some experts who can build you a more sophisticated iteration of what you initially created.

- **Test materials.** The materials are crucial to your product. It seems obvious, but you won't know how many choices exist, such as wood, metal, fabric, plastic, or other source material, until you actually start looking. After one of Mina's prototypes was made in iron, it was immediately obvious that it was too heavy. Getting prototypes in three different materials— aluminum, zinc, and plastic—helped her figure out what would work best in terms of strength, flexibility, weight, and cost.

- **Break it.** The quickest way to find some flaws in your product is to break a couple of prototypes in the material that you will use for final production. Initially, you don't have to go to a product-testing firm to test the capacity of your product, although you will definitely want to before bringing it to market. Mina's initial tests entailed borrowing a couple of kettlebells from her CrossFit gym, hanging them off of her fence, and adding weight gradually. You could also give your product to a friend to test and ask her to break it. Learn from this, and make a better one. Eventually, Mina had her product professionally tested to make sure that the hinges of the Heroclip still worked smoothly after folding and opening 2,000 times, that magnets didn't fall out after being dropped, that the product could hold more than what she advertised, that the metal the factory used was what they said they would use (sadly, this cannot be taken for granted), and that the packaging and finished materials complied with regulations, (such as California Proposition 65, which requires products to have warnings if they contain certain chemicals known to be harmful to people).

- **Make the final prototype.** This one is called the presentation prototype, and it needs to be polished, as you will be showing it to all of your stakeholders (like investors and potential customers). If you had multiple prototype units made by your factory (which we strongly urge you to do) and your most recent round managed to survive your homemade or professional tests, you already have one. To make it, find the people who have the skills you don't. You get what you pay for. If you are making a wooden toy, you don't have to be an expert woodworker; you just need to find one.

- **Make a manufacturing prototype.** Depending on your product, you might need a *manufacturing prototype*, one that can be used to create the tooling used in the manufacturing process. In Mina's case, her metal prototypes were good enough to take to factories to get quotes.

Here's our key message: Mess around with it yourself to get the basics done. Then get professional help.

PATENTING YOUR SNOWFLAKE

One question that comes up as would-be InventHers hone their ideas is: When do I patent my product? We are not attorneys (although, actually, Hilary is a recovering attorney), and you really need an intellectual property specialist—but here are some of our thoughts based on our and others' experiences. Most people are afraid that their amazing idea will be stolen and are very hesitant to get feedback on the idea for that reason. But, ideas are a dime a dozen, and it is actually what you *do* with that idea that makes it valuable. We could have two people with the same idea and end up with two completely different products that do similar things. We are big fans of getting lots of feedback early on so that you can make smart decisions about how to allocate your precious resources of time, energy, and money. However, when you have a working prototype, you might want to start thinking about protecting your intellectual property: you cannot patent a concept, only an invention based on that concept. We suggest that, when you have a working prototype, file a provisional patent application (PPA). This application is not as rigorous as an actual patent but gives you one year to decide whether you want to make the investment of time and money to file for a full patent. While you make up your mind, you can claim "patent pending" on your products (for that one year). A provisional patent costs between $1,000 and $5,000, depending on how much work you will do yourself, while a full patent can exceed $10,000. And there is no guarantee that your patent will be granted. Stop by uspto.gov (the Patent Office website) to find out more about how to file a patent application (also see Resources, page 233), but plan on getting an attorney at some point to protect your investment.

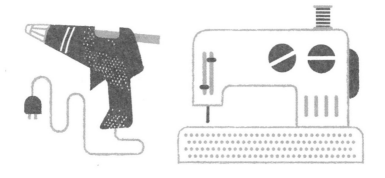

Naming Your Product

Even at the prototype stage, you need something to call your product besides "you know, my product that is like a hook and a carabiner." And it's hard to come up with a name, especially one that isn't already being used for an existing product. Aim high and believe that your product is one day going to be out on the market. Don't confuse anyone from the get-go with a name similar to another product. Mina had a heck of a time coming up with a name, even after many brainstorming sessions with kind (and opinionated) friends. Here are a few of the names she considered and ultimately rejected for what is now Heroclip:

- The Hookr (this sounded really good when Mina's friend Francine came up with it over wine, but not so much the next day)

- Stay Hook (but isn't the product about being active?)

- The Beaner (this sounds vaguely racist . . .)

- Third Hand (sounds obscene)

- The Roo (as in kangaroo; too cutesy and more appropriate for a sack of some kind)

- Glenda (as in the good witch that helps get things done; too girly)

- Freehand (wasn't there some software by this name about a million years ago?)

Mina really couldn't come up with a good name, so she just started calling it a LulaBiner ("Lula" from the legal name of Mina's original company—"Lulabop"—and "Biner" because it is a variation on a carabiner). After Mina got her final prototypes and realized through use, testing, and focus groups that she had a product that could be useful to people beyond hikers and moms, she and her team worked on coming up with a name that was somewhat gender neutral and gave some indication of what the product was. She also wanted a name that could be used throughout the product line (at this point, Mina had already realized that there would be different models of the product). The final pick: the Qlipter (pronounced "Clipter"), and the next product in the line, Qliplet. Two years later, after realizing that customers had difficulty correctly pronouncing and spelling of these names, Mina and her company decided to do some major rebranding and the product became Heroclip, an easy-to-pronounce, easy-to-remember, easy-to-translate name that represents how the product is a sidekick that makes its user a hero in her daily life.

Stephanie Lynn
Sweet Spot Skirts

SEW SUCCESSFUL

Stephanie Lynn was a real estate agent in Vancouver, Washington, in 2009 when she threw an old disco miniskirt over her bike shorts to meet some friends for dinner. Even though she was sweaty and helmeted, her little skirt drew compliments. After it happened a second time, she realized the little-skirt-over-athletic-gear idea had potential. She knew she could create a better, cuter skirt for the active gal, one that would cover the "sweet spot." (Anyone who has been behind a female cyclist knows what she's talking about.) A skirt that would look good on anyone's behind—not just ultraslim athletes, but women like herself (a size 12). Stephanie was not a seamstress or textile artist, but she knew what she wanted. And she needed a prototype.

"First, I drew out a design on tracing paper on the island in my kitchen. Then I bought fabric—that was key to the vision. It's not just the design, it's the material. Then I started searching for a seamstress. I had a high school friend draw up a 'don't steal her idea' release, and I hired one."

Once they had a working pattern, they were able to sew fifty as a first batch, hiring seamstresses from Craigslist, mostly stay-at-home moms, to sew them in small quantities. Lynn branded each one herself: "Labels were expensive, so I made one by heat transferring our logo onto a ribbon. I then ironed them onto all fifty skirts."

Lynn paid fifty dollars to attend a bike swap in Seattle, where her skirts quickly sold out and she caught the attention of other retailers. She was officially in business. When REI came calling, she had to quickly scale production to meet supply demands. Today, Sweet Spot Skirts is one of the Northwest's fastest growing start-ups and shows no sign of slowing down.

But it all began with the prototype. Lynn has a wall in her Vancouver store devoted to her prototypes, where visitors can view the iterations that evolved into the bestselling skirt. She encourages entrepreneurs not to be daunted by the process.

"You have the idea," she says. "Then find someone with the skills you need to make it reality."

Sirena Rolfe
Tempus Hood

PROTECTING THAT PROPERTY WITH PATENTS

She did it for the do. The hairdo. Sirena Rolfe, like many women, spent good money and time on her hair, only to be frustrated when rain ruined it. Umbrellas weren't the answer—too many people lacked umbrella etiquette, and she was tired of being poked and dripped on when strolling in urban environments. Plus, who wants to carry that thing all day? And for places like her home in Colorado, where all four seasons and all types of weather can appear in a single day, she wanted a better solution. As an outdoorswoman, she wanted to be out in all weather but didn't always want a bulky coat just to have the benefits of the hood. Her women friends agreed—there was a need for a new solution. So in 2012, she sketched out an idea: a packable hood that could attach (or not!) to any jacket, keeping hair protected from the elements without the extra baggage.

Like all InventHers, Sirena had a prototype made and found a manufacturer. For her product, she chose a local seamstress who could make her hoods in small batches with maximum flexibility for colors. She's proud to have the "Made in USA" designation, as well as "Made in Colorado."

What is her greatest asset? The idea. The intellectual property that is the Tempus Hood. Sirena went forward to protect this idea by filing for a patent, which was just approved in 2019 after a five-year process. She states that she is one of the very few African American women (since the late 1800s) to be the sole owner on a patent. "I did some research, and unfortunately the published data isn't consistent, but I am in the top ten," she said. "I'm ecstatic to be one of them."

Getting a patent requires an attorney, and it can be expensive, but Sirena cut costs by doing a lot of early legwork herself. This doubled as market research, as she searched the Patent Office database for any products that might be similar to her own. (An attorney will charge for this.) First, she filed a provisional application—this gives you a year to finalize the idea. You must file for a patent by the end of that year, but in the meantime, it gives the Patent Office (and others!) notice that it's in the works. Then she interviewed attorneys. The first one acted like he was doing her a favor even by meeting with her. The second one was just looking for one more patent before retiring. The third was excited by her idea and had a background in product engineering. He got it—and her business. Together, they got to work and got her patent filed and approved.

It cost Sirena approximately $12,000 and five long years to get her patent. But the business is growing, and Tempus Hood is a protectable asset. Her goal is to license the hood to a larger outdoor company that has greater manufacturing capabilities and can bring it to an even larger market. She's already making history as one of the first female African American patent holders. We expect to see her in the books as a billionaire tycoon soon—and her hair will look fabulous, thanks to the Tempus Hood.

Chapter 3: Customer Identification
Finding Your Peeps

So you've got an idea that just won't leave your head and that your gut is saying is a good one. You've Googled to check out potential competition. You've Googled to learn about the general condition of the industry your product will be in. You've mustered up the courage to get some honest feedback from your First Five. You even have a prototype. You're on cloud nine—your invention is going to be *the next big thing!* Before you start planning your outfit for the cover of *Entrepreneur* magazine and planning a celebration for you and your new best friend Sara Blakely of Spanx, let's figure out if people will actually pay for your product, and if so, how many they will buy, and how much will they pay, and . . . and . . . .

Yes, there's a bit more work to do.

We'll put it bluntly: Even the very best, most useful products in the world are not going to appeal to everyone. Or in some cases, *anyone*. Even if Google tells you that your product has *zero* competition, that doesn't mean your invention will sell. We know it has never crossed your mind, and we hate to sully your beautiful thoughts with realism, but there is the dreaded possibility that no competition exists because no one wants a product like the idea in your head, at least right now.

In the happy scenario in which people *are* willing to pay for your invention, there is the issue of whether *enough* people pay money to make your idea worth your time. What is a perfect idea or invention for you might not resonate with a large enough audience to make it profitable (you *were* looking to make money with your product rather than to have an expensive, exhausting, and life-consuming hobby, right?). It's time to let your idea out of your head and get some input from real people to find out who the right customers are for you—it's time for some customer-centered market research.

We know that "market research" sounds corporate (and possibly boring), but it is just a fancy term for learning about your potential customers and their likes and dislikes. (You've already done some initial market research by looking at the potential competitors out there.) Large companies spend tens of millions of dollars on it every year, but trust us, you don't need all that. Here's what you *will* need: thick skin (you may not always hear what you want to hear), spunk (you will definitely be talking to a lot of people), and willingness to adapt your plans based on what you find. This chapter's InventHer stories are about using the results of market research to find your ideal customer base. As InventHers, we need to always have open minds about whatever the market is telling us and respond accordingly. (Mina's husband, who is a student of world economics, says, "Consumers are never wrong." We have to agree.)

Market research is all about asking some basic but hard questions. (This is the one area where we actually will sound a bit like those MBA textbooks, and that's because answering these questions is nonnegotiable for anyone attempting to turn their invention into a business.)

1. **Who** are the customers who will buy my product?

2. **Why** will they buy my product?

3. **How much** are they willing to pay for it?

4. **Where** will they hear about it and buy it?

Although we are going to talk about each of these separately, the who, why, and how much all intersect and affect each other. The where is all about marketing and distribution, which will be discussed in depth in the next chapters. If your customer is a working mom, you wouldn't expect her to find your product through a morning talk show when she is presumably at work, right? Similarly, if you need to price your new exercise machine at $10,000, your market is probably not the average US household. Consider these important elements together as you figure out who your ideal customer is.

Who?

Who? is obviously the most critical question and the starting point of figuring out your customer. The who has to be narrow enough for you to really get to know them so you can *talk to* and *reach* them in a way that resonates with them, but large enough for you to make the kind of sales volume you want.

Narrowing down the *who* is surprisingly difficult. Most inventors think they have a general idea of who will be the customers for their product. But dig deeper. It's not enough to say it's for busy moms, or millennials, or teenage boys, or worse—the highly optimistic "It's for everyone!" (We've yet to see a product that is truly for everyone.) Maybe you are making a travel product—for example, the new type of airplane pillow we brought up earlier. You might look at the entire market for travel goods. Some quick work with your browser might tell you that the travel goods industry is a $30 billion market.[5] Sounds great! But is your product more geared to a business traveler or a leisure traveler? Is this someone who travels once a year or once a week? Is it for men or women? People traveling first class or economy? Identifying your customer means drilling down and seeing how big this particular population is, and also whether this population can afford your product.

Big companies with huge marketing resources spend time putting together "personas"—composite sketches of segments of their buying audience. They will give them names like Sally Soccer Mom and assign job titles, age, gender, marital status, education level, challenges, etc. They'll even decide what they read, what their hobbies are, where they travel . . . you get the idea. In our opinion, sometimes big marketing teams can get lost in the minutiae of these personas, developing fictional people past the point of usefulness.

 HOT TIP: Your ideal customer may be different from what you initially thought.

As a scrappy InventHer, there is no need to get into this depth. However, the theory behind the persona exercise is valid. It's all about *segmentation* of your market. Market segmentation is just a fancy way of dividing your customers into smaller groups, typically based on behavior, demographics, geography, and psychography (how they think—values, attitudes, interests, and lifestyle). While it isn't necessary to create elaborate personas, you should definitely think about market segmentation—it's key to determining whether you have a large enough market and, if so, how to attract them to your product. With our travel pillow example, you may decide that it is perfect for parents of infants rather than business travelers. With this realization comes a slew of other insights. For example, even if these parents *need* and *want* the product, will they actually buy it, and if they do, where will they buy it? Having both been moms of infants, Hilary and Mina know that when you are traveling with infants, the last thing you need is another item to carry, no matter how much you want it. And they also know that with infants, the airport store is probably not in your cards (unless you enjoy seeing merchandise arrangements in tight quarters come tumbling down thanks to your infant and the five bags you are juggling). So . . . maybe parents of infants will want your product the most, but maybe they are not necessarily the ones who will buy it the most often, especially if your idea is to sell it at an airport store. Being as specific as you possibly can, or segmenting the market, helps you identify who is your most low-hanging customer.

Determining the *who* is more than a mental (and Google) exercise while sitting around a conference (or dining) table with a whiteboard (or the back of your utility bill). Your brain, your experiences, and your knowledge of who you think your customers are will give you a *hypothesis* about the ideal customer for your invention. But then, like any good researcher, you must test this hypothesis to see if you are right. And you do this by going out and getting input from real people. Large companies take this to one end of the spectrum, hiring ethnographers to shadow their hypothesized customer for days and weeks to record exactly what she does, where she goes, whom she talks to, what her daily challenges are, what she buys, what she eats, and so on. An InventHer can't always afford to take a month off from work and family, so later in the chapter we provide several different ways you can get valid input that will guide you in reaching your customers and convincing them of the benefits your product will bring to their lives.

Here's a real-life example from Mina's Heroclip experience that illustrates the importance of market segmentation. Heroclip can technically be used by anyone, but as a small business with a matching small budget, Mina and her team couldn't possibly try to sell to everyone. Should it be for the outdoors lover carrying a lot of adventure gear? DIYers with a paint bucket, a tool belt, and five other things they absolutely need to fix their wall? Fashionistas who don't want to put their $1,000 purses on the ground at the newest, hippest restaurant? Or college students with their fifty-pound book bags?

In the end, Mina and her team decided they would focus on the outdoors market. But then they realized how diverse it was. There is the Everest-climbing outdoors lover who thinks nothing of spending hundreds of dollars on gear. There is also the more competitive outdoors lover (think trail running) and the casual outdoors lover (think a slow walk around a neighborhood path). Each of these groups has their own buying habits (where, when, how they buy) and what they think is a reasonable price for gear like Heroclip. This is when company values and philosophy came in. Although the practicalities of who will buy your product are important, market segmentation and homing in on your key customer can also depend on your values. All things being equal, who do you *want* to sell to? What kind of person's values align with yours? In the end, Heroclip's priorities are about getting the average person (not the Mount Everest types) active and outside, and these priorities affected what they ultimately decided was their target market. Making the decision to target casual outdoor enthusiasts affected pretty much everything at Heroclip, from the way they wrote their product description to the pictures and imagery they used, the features they highlighted, their price point, where they advertised, and what media they targeted.

 HOT TIP: You need to figure out who your target audience is, based not only on your product's uses and its benefits, but also your company's values. Take feedback, and be willing to shift from original assumptions.

Why?

As you prepare to get real-world feedback from your potential customers, an important question to ask yourself is why your customer will buy your product. Five people may buy your product, but it could be for five completely different reasons. Figuring out what those reasons are will affect how you talk about your product and how you ultimately reach your customer. It might also require you to make some hard choices about whether you want your product to be used in a particular way (you don't want someone using your invention for nefarious purposes, do you?). Asking why a customer finds your product valuable can also yield unexpected opportunities. With Heroclip, as her team grew to be able to explore markets beyond their initial target customer, Mina learned that the item she had invented as an outdoor and family product was being used quite a bit by DIYers and building contractors for easy access to tools. This led the Heroclip team to expand into the hardware market, eventually landing the product line at Ace Hardware stores.

How Much?

As we know as consumers, one of the factors that affects whether we want—and will pay for—a product is how much it costs. Hilary once consulted for a company that was developing a supercool electronic accessory for teens. Early market research from a focus group was positive. They all loved the idea and said they would buy it. The one downer question: How much would they pay for this fab new accessory? She conducted a survey with a free online tool. The vast majority of respondents said they would pay about twenty to twenty-five dollars. Bad news for a company that was looking to sell it for about $100. They ignored that, going on the assumption that the end product would be so great customers would pay more. Spoiler: they didn't. The lesson here is not that nobody wanted the product or even that the price was just too high, but that the *combination* of the price and the target market was off. We also think it is seriously wrong that the company completely ignored what their respondents said. Always take your customers' (or potential customers') input seriously, and if you decide you are going to ignore what they say, have a compelling reason (although we ourselves cannot think of a single good reason for ignoring what potential customers say—at the very least, we will discover that they are not our customers).

Here's the thing about pricing. It is not as simple as "lower is better." In fact, pricing can be very complex and requires taking into account many variables (like value perception and consumer psychology) that keep a lot of pricing strategists in business. When you are just launching the product, however, three crucial things highlight the importance of knowing your customer.

First, the value your customer gets has to justify the price, but value doesn't just come from the functionality of the product. Value could be something intangible like status or doing good or some reason completely unrelated to the features and benefits of the product itself. For example, the function of a Louis Vuitton handbag is to hold stuff, the same as a two-dollar shopping bag, but people pay thousands of dollars for a Louis Vuitton purse because of the status it conveys. Another example is Toms shoes, which might look like a much cheaper shoe but can justify a higher price because part of that cost is helping kids in developing countries. What does your customer care about, and is she willing to pay for it?

Second, your pricing should be reasonable within your product's category. Unfortunately, even if your product can do what ten products can, it doesn't mean you can price it ten times higher than a competitive product in the same category. People have set perceptions of what things should cost, and this guides their decision making. For example, people might think twenty dollars for some nails and wire in the hardware section of a store is ridiculous, but in the art or picture frame section, boxed up nicely, twenty dollars may seem perfectly reasonable.

Third, the price you want for your product and the price a customer is willing to pay have to line up with the price you pay for your product. This is something that seems so obvious yet is overlooked time and time again. This is why, in the prototyping chapter, we emphasized that you should start your product development process with *manufacturing in mind*. In the case of Hilary's aforementioned client, even if they had come down to earth and decided to price their product at what their customers would pay, twenty to twenty-five dollars, this would not have even covered the cost of manufacturing. Do your pricing research first, before you make the same mistake they did: spending millions on product development only to have a product that no one wanted at the price they had to charge. Having said all of the above, once you launch your product, if you are selling directly to your customers on your website, don't be afraid to experiment with small variations in pricing. A nickel difference in your pricing can make a huge impact. Case in point: with Heroclip, changing the price from $20 to $19.95 tripled sales.

A Final Word

To summarize, make sure all of your who, why, and how much answers align with each other. We have seen too many products that claimed to be for one type of customer but were priced for a totally different one. For example, if you determine your travel pillow is for college students studying abroad, your differentiator is that it's the only one made with a premium sustainable cashmere, and the price point is in the very expensive range, you've got a problem. A study of your segmented audience (students) will show they don't have the disposable income for expensive goods. The who, why, and how much have to be in sync and will affect the where, as you will see in Chapter 4: Distribution (page 65).

Doing Your Own Market Research

Here is your guide on conducting research to test your hypotheses about who your target customer is, why you think they will buy your product, and how much they will pay.

QUESTIONS TO ANSWER

The kind of questions you ask will determine whether you get useful answers that can really move your invention forward or . . . well, totally useless ones. In Chapter 1: The Initial Idea, we asked the First Five for general feedback on the merits of the product. This time, we are going to dive deeper to determine whether the people you think will be your customers actually will be your customers. Here are the questions you are trying to answer:

- Who needs this?

- Who will simply want this?

- Why will they want this?

- Where will they want to buy this?

- How much would the people who need this be willing to pay?

Now that you know the questions, how do you find the answers? There are lots of ways to do this. Let's look at a few, starting from the easiest to the most in-depth.

Friends and Family Opinions

Go for the low-hanging fruit. You've already talked to your First Five. Talk to your friends and family and ask what they think. Be aware that these folks will probably (though not always!) offer the gentlest opinions. If you've been thinking about this product for weeks, loved ones can offer up valuable input: sometimes you can't see the forest for the trees and are missing something obvious. Reminder: kids are brutal. They don't hold back.

Professional Opinions

Developing a piece of fitness equipment? Find a professional trainer or other fitness guru and ask what she thinks. Selling a pet accessory? Find a veterinarian and seek her opinion. A new kitchen tool? Find a chef. You get the idea. Find people who would be more than casual consumers, and see what they think.

Surveys

There are lots of tools like SurveyMonkey or Google Surveys to gather feedback. Even a Facebook or Twitter poll can provide some data. The difficult part here is getting a diverse enough sample and giving a sufficient incentive to fill out the poll. Your friends and family might do it, but you need more. When sending out requests to strangers or unknowns, it is common to offer an incentive, like a gift card or a chance to win something. However, some people will just do it for the incentive and won't take the time to give productive feedback. That said, a survey sent to an appropriate professional group with a heartfelt "Hey, I would really appreciate feedback on my new product!" can yield some terrific results. Think about Facebook groups, coworkers, LinkedIn groups, professional organizations, alumni groups—any group you have a connection with might be a good possibility.

Crowdfunding

Where would we be without Kickstarter and Indiegogo? These crowdfunding platforms are a great way to test a product's market potential. The downside: you need to have at least a basic prototype to shoot a quick video and show your product to the world. You can set a fund-raising goal, and if you don't make it, you don't have to produce the product. Either backers contribute money to help produce your product, or they don't. Be aware that a crowdfunding campaign is a big endeavor, but since backers actually put money where their mouths are, rather than merely declaring that they *would* buy the product, the feedback you get is incredibly useful. (Learn more about crowdfunding in Chapter 7: Financing Options, page 149.)

Small Launches

What better way to test your market than to start making and selling your invention? This only works if you can readily produce small batches of high-quality product. Many successful projects have started with an Etsy shop or a small display in a local retail store. The key is to have a way to get feedback from customers—either by talking to them directly or by getting reviews on your Etsy shop. There are lots of ways to sell small batches directly—we know entrepreneurs who got their start at farmers' markets, craft fairs, and church bazaars (Martha Stewart started by selling pies at a stand in front of a local store). This approach is definitely better for folks who are making something they can manufacture themselves—apparel or crafts come to mind—rather than hard goods that need special facilities to produce.

Focus Groups

If you really want to delve deep into your potential market, consider a focus group. Don't be intimidated by the marketing jargon—it's just a bunch of different people that you're soliciting consumer opinions from. Include some family and friends, but find a range of people from different backgrounds and demographics. They will all come together for a meeting where a facilitator (it can be you, but an experienced person is even better) asks questions about the product, and you can hear from the participants while they hear from each other. Unlike a survey, where people are answering questions in a vacuum, a focus group thrives on interaction and gets people responding to other participants' answers. Find some folks and see what wisdom the group has for your product!

Sometimes Your Market Finds You

Giving yourself time to explore your market sometimes reveals a different market than what you've anticipated. We know one InventHer who started a great business selling rain skirts—waterproof, wraparound overskirts to keep your legs dry in those spots a raincoat fails to cover. She was a nurse who was tired of going to work with wet pants after standing out waiting for the bus (yes, she is from Seattle). She thought she had a huge hit with dog walkers and other folks who had to go out in wet weather and didn't have hands free for an umbrella. She started selling them on her website. Guess who discovered them and found them a great fit? Monks. Her skirts turned out to be a perfect solution to wet monastic robes, and she had some initial brisk sales once they spread the word to others. However, taking a look at the entire monastic consumer market, one can quickly see that it probably wouldn't be big enough to sustain her product. In the end, that InventHer decided to move on. Knowing when to fold is also part of being an InventHer. There is no shame in discovering your market isn't quite there—take your knowledge and move on to your next big idea.

Don't be afraid to be proven wrong. We've seen businesses fail before they get off the ground because their attempts at market research didn't validate what they thought was their audience and they ignored the evidence. Market research isn't just a checklist item on your journey to being the next great InventHer. Listen to your customers and make decisions accordingly. Customers are the only judges who matter.

Michele Mehl

Excy

UNCOVERING YOUR MARKET

Michele Mehl never set out to make a traditional exercise product. She created Excy, a portable exercise cycle / machine that turns any chair into an exercise bike and can be moved around the home or office to get a workout wherever you are. "It was never about losing weight, looking good in a bikini, or building muscles. It was always about health and wellness," she said. The original inspiration occurred when she was at the park, watching her young son, Jack, play. She was sitting on a park bench, thinking that she could be using that time to exercise, as she had just turned forty and was not happy with her current state of fitness. Stepping up and down on the bench was boring. What if any bench could be an exercise bike? She called her genius uncle, who had experience making products, and soon the idea for Excy was born. Their planned market: Excy would be for any busy mom or parent who didn't have time for exercise. It could be used under a desk, in front of a television, anywhere. They got busy on prototypes.

During the prototyping phase, there was a life-changing development. Michele broke her leg—badly. It was a serious break that required multiple surgeries and ten screws in the leg. Then she developed a blood clot while she was in the hospital—a common occurrence tied to lack of movement. She started to use Excy for upper-body workouts while her leg healed. Mehl suddenly saw the potential of Excy for physical therapy, as she embarked on her own PT and rehabilitation. "Getting hurt shifted everything," she said. "You start noticing others around you with limps and disabilities. Excy wasn't going to be just for busy people—it was for people who were really struggling with disabilities and rehab. I created an entire rehab facility that goes where you go."

Her first test market, a Kickstarter campaign, bore out her findings. "I talked to our customers and found out who was buying our product," she said. "They found us—people with disabilities, injuries, and health issues. That led to one of my most pivotal moments as an inventor. I brought Excy to a group of special needs children, and a girl with Down syndrome loved it. She said, 'I can do just what my brother can do!' It was extremely exciting. Excy wasn't going to be for the soccer mom on the side of the field—it was for people who were really struggling."

Today, Excy's market is at the thriving intersection of the $30 billion US consumer health and fitness market[6] and the $29.6 billion outpatient rehabilitation market.[7] It's currently being used in Stanford Hospital, where patients are exercising in their hospital rooms with Excy. Mehl has found her customers—they just weren't who she expected.

Fran Dunaway
TomboyX

THE ART OF THE PIVOT

Fran Dunaway just wanted a nice shirt. A nice button-down, men's-style shirt, made from quality fabric, with some interesting details around the collar and cuff. Maybe a blazer and a polo-type shirt too. Things that might be traditionally men's styling, but fit for women. With her wife and cofounder, Naomi, she came up with a cute name, TomboyX, and they bootstrapped their way to some product and launched a successful Kickstarter campaign. That was in 2013. But a funny thing happened as they began this new brand, targeted at LGBTQIA women like themselves. The customers came, immediately identifying with the TomboyX name, and started talking. They didn't see themselves represented in traditional clothing ads or marketing.

They were clamoring to be seen and to have an inclusive brand of their own. They started making requests, and the most common was "Could you make boxer briefs for women?" Fran and Naomi went online to search for these, but all their searches revealed was Spanx. They set out to make some great underwear, from size XS to 4X. After some research, they realized they needed $14,000 to get a line of underwear out under their brand. They didn't have the cash in the bank to fund this (for more on this, see Chapter 7: Financing Options, page 149), so they did an online presale to test the waters. It sold out in two weeks. Fast-forward to 2016: less than three years after that initial Kickstarter campaign for a shirt, TomboyX pivoted to exclusively selling underwear in their unique style, for women (or anyone), and in any size. It's working. The company recently closed a Series A round of financing, and the quality of their underwear keeps winning fans.

TomboyX's website states, "We have an agenda. It's not a gay agenda. It's not a feminist agenda. It's not a butch or fem or alphabet soup of letters agenda. It's all of those things. It's a human agenda. An agenda that says, 'We are all people.'" Their agenda and inclusive message, along with their openness to customer feedback, has made them a near cult favorite. Fran and Naomi were able to pivot on the actual product, from shirts and blazers to underwear. They listened to their well-defined market and filled a need. Defining your market is a critical part of market research; listening to it is another. Be the listener.

Chapter 4: Distribution
Getting Your Stuff Out There

One of the most glorious feelings in the world is seeing the product that came out of your genius head in the hands of a customer (bonus giddiness if you catch her being an inadvertent salesperson and explaining all the awesome features of your product to her friends). Back in the day, to buy something, a customer actually had to get behind the wheel and drive to a local mall (or beg for a ride if you started shopping young like we did). Nowadays, anything we want, we can find fifty different ways to buy it. Great, right?

Well, yes, and no. What's great is that you can pretty much start selling as soon as you have something to sell. There are two things that aren't so great. First, if you try to sell through all of the different ways there are to sell, you will probably pass out from sheer exhaustion or have a heart attack from overdosing on caffeine. Second, no one knows about you or your product, and with so many ways to buy, it is virtually impossible for someone to "bump into" your product. We're here to help.

There are many ways to get a product from the maker (you) to the eagerly waiting hands of the customer. Each method has pros and cons, and it is up to you to decide how you want to distribute, depending on the nature of your product, your resources, how quickly you need to start making money and receive payment, and so on. Let's start with some basic definitions.

Distribution means the way in which you get something from Person A (you) to Person B (and C and D and E and so on—hopefully you will quickly run out of letters to designate your customer!). A distribution channel consists of all the different hands a product has to pass through as it gets from a manufacturer to the end user. And yes, we used the term *end user* because depending on which distribution strategy you choose, your customer is not always the one who will be using your product. This is the case if you sell a boatload of product to Costco. Your customer is your buyer at Costco, but the end user is the person who buys and then uses your product. We can make it a bit more complicated and say that the person who buys it from Costco may not even be the end user if she is getting it for someone else, like her kid or her team members. Another term for end user is *consumer*, or the person who, well, consumes, or uses the product.

We'll look at these distribution channels from the shortest to longest.

Direct to Consumer (DTC)

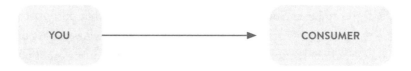

The simplest distribution channel is one where the maker of the product sells directly to the consumer. A commonly seen example is when someone who handcrafts jewelry (or makes a delicious pie or jam) goes to a farmers' market and sells it. Or when someone makes pies and sells them right out of her home. A more modern example would be selling directly to a consumer through the product's website (speaking of which, check out the websites of all the InventHers who contributed their stories in Resources, page 233.)

Notice we said a *product's*, not a retailer's, website. When you buy from Target.com or Walmart.com, even though you are getting the product directly from a website purchase, you are buying from a retailer, not the manufacturer, so this would not be a DTC experience (more on other sorts of distribution later in the chapter).

One of the first things you should do when you have a product is to make a website (come on, you already knew that nothing really exists until it lives on the website, right?). Not only is a website a place where you can sell, it is where you can really take your time explaining the features and benefits of your product—a website is like an infomercial where you don't have a time limit.

So, how do you get started on a website? You can spend tens and hundreds of thousands (or maybe even millions?) of dollars getting "experts" to make your website, but you can also use the many platforms out there that make it super easy to create simple websites quickly. Hilary used Squarespace to build her own website, while Mina used Shopify to get her product store going (in fact, she still uses Shopify). There are many others that purport to be "plug-and-play" (or "drag-and-drop"). You can review our list of some commonly used website platforms in the Resources section (page 233), but new platforms are popping up every day, so make sure you do a full internet search. Most of these platforms charge monthly fees based on how many orders you process each month. Since you are just starting out and won't have many sales initially, the good news is that your monthly fee will also be low (we are silver-lining type of girls).

Here are two basic features that most e-commerce platforms provide:

- Some kind of template for the main pages that any e-commerce business might need, like a home page, a collections page, a product page, a blog page, and other content pages. A home page is where a customer lands when they type in your website address. A collections page has the different product lines you offer (for example, the Heroclip line and branded merchandise). The product page is where you provide descriptions of your product and where customers can click on that lovely Add-to-Cart button that leads to a sale. With templates, it is relatively painless for someone with a bit of computer experience to put together a basic website—you just drop in your own images and text to make the website all your own.

- Most platforms also come with a built-in payment processor so that your customer's credit card can be charged and the money deposited to your bank securely. Different platforms have different rates they charge for credit card processing (often this depends on volume), but generally there is a fixed component (like thirty cents per transaction) plus a variable component (like 2.9 percent of the order amount). So, make sure you look into fees like these when you are selecting a platform.

Note that social media platforms like Instagram, Facebook, and Pinterest also offer the ability to sell products directly, but your website is going to be the mother ship where you can control the conversation and talk about your product exactly how you want. And don't forget that you can have the most beautiful, most informative, most compelling website, but if you don't get traffic your bank account will pretty much sit at zero. Make sure you read Chapter Five: Marketing (page 93) and shout about your product from the top of your lungs (we're speaking metaphorically, of course, although we would give you a high five if you literally shouted about your product in a big crowd of potential customers—moxie goes a long way).

Best Practices for Your First Website

You can go cheap and fast when building your website (basically creating your online infomercial), but there are a few rules and best practices that we and our InventHer friends have learned along the way.

- Take great product photos, with an uncluttered background. One mistake Mina made early on was that she was so excited about showing all the great things people could do with a Heroclip her website barely contained any images of the product itself, which very quickly led to a full inbox from customers who couldn't quite tell what the product looked like. In showcasing your product, make sure you are not engaging in any photographic, lighting, filtering, or photoshopping wizardry that makes your product look different from what it actually is. One of the worst things you can do as the seller of a product is to deliver something that is not what the customer expected, and you do not want to deal with all the correspondence (or worse, returns) because your photos did not accurately represent your product.

- Go overboard pointing out features. Since you have spent so much time with your product, you might think that all of the cool things that make your product unique are obvious, but we have learned to take nothing for granted, especially when it comes to customers' knowledge of a product. Point out the main features over and over and over.

- Along the lines of not taking anything for granted, make sure you clearly articulate (maybe multiple times) how to use your product. A video (even a homemade one) goes a long way, even if you think that your product is intuitive and its use obvious.

- Make sure you have a detailed FAQ (frequently asked questions) section to which you can point your customers. Even if you put every detail of the product on your page, your customers may skim over them. If you have an FAQ page, they will go there before contacting you (and if they do contact you with a routine question, you can say, "Great question! Please find the information you are looking for on our FAQ page," and include a link to the FAQ page).

- Make sure your website works quickly, especially on mobile devices. According to research by Google, one out of two visitors to a website expects the page to load in 2 seconds, and 53 percent of visitors will leave a site if it takes longer than 3 seconds to load. Yes, the world keeps on getting faster and faster, as are consumers' expectations of their shopping experience.[8] We live in a speedy world, and if we want to keep our customers from moving on to the next thing, we need to have our websites zipping along.

These tips are universal best practices, but as always, there are many ways to attack the beast.

Third-Party Marketplace

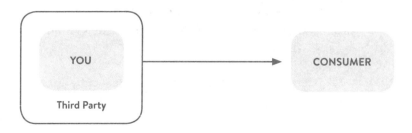

"What about Amazon?" you ask. What? What's Amazon? Just kidding. Unless you have just time traveled from the early 1990s, you know Amazon is something that is always at your fingertips for quick gratification of whatever you need or want. Amazon and other third-party marketplaces (like Etsy and eBay) allow you to create a listing on their site and become part of the marketplace for a fee or cut. The appeal of these marketplaces is that they already have a large (gigantic, in some cases) customer base ready to buy. The difference between selling through a third-party marketplace and a retailer like Target is the relationship between you and the consumer. On a third-party marketplace, the consumer is still buying directly from you, but within the confines of that particular market vendor (think of a farmers' market—you still buy directly from the farmers, but you find them at that farmers' market, not at their farm).

So . . . is it worth the fee to sell your product on this massive marketplace filled with customers poised to click on the big yellow Add-to-Cart button? From our experience, Amazon is such a convenient way for customers to shop that selling on it is pretty much unavoidable. During the thirty-six hours of Prime Day sales in 2018,

Amazon sold 100 million items—that's 771 products sold in the time it takes you to say "One Mississippi."[9] With this kind of sales traffic, even large brands and retailers are listing their products on Amazon in addition to their own websites.

Some advantages of Amazon (aside from its massive size):

- On Amazon, you can either ship directly to the customer (called merchant-fulfilled) or use its warehouse and fulfillment services (known as FBA, or fulfilled by Amazon). Using FBA services involves shipping large quantities of your product to one or several of Amazon's warehouses. Whenever someone makes a purchase, an Amazon warehouse worker ships it to the customer. You will pay for all of this (including a warehousing fee, per-order fee, a pick-and-pack fee, and postage), but streamlining your operations may be worth the cost.

- Because of its volume, Amazon can ship things quickly at a much lower rate than you can. So, if your customer has Amazon Prime and you choose to have your goods fulfilled by Amazon, she will get your product in one or two days (or sometimes even on the same day). So many people are used to receiving purchases immediately that shipping speed can be a serious draw.

- Many, many, many people already have their Amazon accounts set up so that purchasing there requires a mere tap of the Buy-Now button, whereas on your website, customers would have to go through the tedious task of entering their names and addresses and credit card information. At Hero-clip, Mina sees quite a significant percentage of customers who will come and learn about the product on the website and then go buy it at Amazon, simply because of convenience.

- Amazon has some cool features and tools that increase the visibility of your product. Besides the ability to have your product listing pop up when people enter particular search terms, you can advertise on other people's listings. It's true! You can literally direct Amazon to show an ad of your product when people are looking at a compatible, complementary, or competing product. Super cool, right? Besides paid ads, Amazon will do things like show items that are "frequently bought together," which can give your product some additional visibility.

And now, the downsides . . .

- The number-one downside to Amazon is lack of control. Once your product has a listing, *anyone* can sell the same product under your listing. Even if you have a totally unique product and are the only one producing it, through various means (see Wholesale to Retailer, page 76), others may get their hands on it and resell it in the world's largest marketplace. Since Amazon customers don't really care from whom they buy a product, multiple people selling the same product will frequently lead to a price war, in which each party will try to get the sale by reducing their price. This is definitely not good for you, and it is critical that you keep a close eye on who is selling your product so you can report unauthorized sellers to Amazon as you discover them. Unfortunately, this is time consuming and sometimes fruitless (although in Amazon's defense, it does seem to be taking this problem more and more seriously).

- The second disadvantage is the lack of flexibility in your listing. Unlike your website, which is like an empty room that can be lovingly filled with beautiful furniture and artwork arranged any way you want, Amazon tells you exactly where to put your couch, how big it should be, and sometimes even what it should look like (Amazon has the right to remove images, which is generally a good thing since this is supposed to prevent sellers from using misleading images). There are also restrictions, such as only being able to list your product in one category. For products that fall into multiple categories (Heroclip fits into "sporting," "hardware," "parenting," and "fashion"), this can be a frustrating policy, since being listed in multiple categories would obviously increase the visibility of your product.

- Navigating the back end of Amazon is not always intuitive, and making a single change on your listing can require endless back-and-forth correspondence with their customer support folks. And even after you've celebrated finally implementing your change with a midday glass of champagne, you could wake up the next morning to see the change reverted. As we said before, lack of control is a *huge* downside to Amazon.

All in all, despite the constraints we've listed and based on our experience and observations, Amazon is just too big a marketplace for most businesses to bypass. And because you can respond to customers' questions and comments directly, you still *can* in many ways affect how your customer experiences your product and your brand. The back-end complications of Amazon are real, though. But luckily for those of you who do not have the bandwidth to learn all its ins and outs, there are myriad agencies that specialize in running the Amazon arm of businesses, and they often only command a small cash retainer, supplemented by a percentage of the revenues they generate. The silver lining of the constraints and structure imposed by Amazon is that Amazon selling is probably the most friction-free channel to outsource, and we think InventHers should seriously consider outsourcing the Amazon channel.

 HOT TIP: Ignore Amazon at your own peril.

Wholesale to Retailer

If the advantage of selling DTC is that you get to interact with the customers directly (even through something like an online marketplace) and generally net higher margins (see A Final Word, page 86), there are also unique advantages of selling wholesale, which basically means that you sell in bulk to a retailer who then sells to their customers. Retailers range from nationwide (or worldwide) giants like Target and Costco to smaller independently owned stores or stores that are completely online (like Zulily.com or Backcountry.com).

The main advantage to selling wholesale is that while making fewer transactions (to a middleman retailer), you get to reach more people to whom your retailer will market and sell. (Despite online being all the rage, did you know that 90 percent of consumer purchases still happen at brick-and-mortar stores?)[10] For example, you sell 100 units of your product to Awesome Store on Main St. in a single transaction (for which you are paid by the retailer at an agreed-upon time); then Awesome Store advertises and sells your product to the customers that come to their store (or, better yet, uses your product to draw people to the store), and Awesome Store handles all the transactions to sell the 100 units. When Awesome Store is close to selling the 100 units, it will place another order with you to restock their inventory of your product (hopefully for more than 100 units after they see how quickly they are selling). If Awesome Store is not able to sell through your product for some strange reason, that would be, to put it bluntly, Awesome Store's problem (but obviously this also means Awesome Store won't be reordering from you anytime soon).

This was a simple example, but, not surprisingly, in real life there are numerous variations, negotiations, and qualifications on the wholesale model that make selling this way a tad more complex than selling to Awesome Store. Here are some things you'll face when selling wholesale to retailers.

The first thing is it is *extremely difficult* to get a retailer to purchase from you (let alone return your call/email/text/letter) as a newcomer with one product (which is what we all start out with). Sadly, this is just the way it is. The good news is that it is difficult, but not impossible. Let's break down the must-knows of wholesale and discuss strategies for wholesale success.

Here's the thing. It might be obvious, but there's one person you need to convince when you are trying to get a retailer to buy your product, and that person is the retailer's buyer. If the retailer is a small, independently owned business, the buyer may also be the owner. If the retailer is a large chain, the buyer (and her team) will be making decisions for a big chunk of the locations, possibly even all of them.

Clearly, the buyer is pretty important, and it can be difficult to be taken seriously by a buyer when you are a small business.

First and foremost, buyers have a ton of people trying to get their products in the door, and in recent years, as retailers reduce personnel due to financial pressures, the buyers have only become busier.

Secondly, when buyers make time to see someone and consider her product, they want to see an entire line of products that they can pick and choose from—generally not just *one* product. Think about it: for the same time it takes to see you and your one product, they could see someone with one hundred products.

Third, it has been our experience that buyers are not always gung ho on being the first to jump on a brand-new product. Buyers are rewarded based on how well the items they bought do, and a new product simply doesn't have the track record to convince them that it will sell at their store. (For more on this, see Selling to a Retailer with Little to No Track Record of Sales on page 79.) One of our advisers, a veteran entrepreneur who has placed items at pretty much every major retailer in the United States, told us, "Buyers don't want to be the first to take your product, and they don't want to be the last."

Fourth, even if the buyer loves your product and is willing to give it a chance, she may be worried that as a new business, you do not have the capacity to meet their demand. Imagine getting a big order and having to say you can't fulfill it because you don't have your production process running smoothly enough (this happened to Mina when she first got her product into a major retailer—it was not a good feeling). Or even worse, saying yes to an order and not delivering. If this dreaded scenario happens, your buyer will be in a major bind because she and the merchandising team have already

Selling to a Retailer with Little to No Track Record of Sales

So what *can* you do as a new business with a new product that has no track record? The first thing is to do all you can to get some kind of traction, whether it is with crowdfunding success or sales on your DTC site, before you approach the buyer. When Heroclip first approached REI, even though they were pitching a major retailer for the first time, they were able to say, "Hey, we sold $400,000 worth of product to crowdfunding backers who loved our product so much that they were willing to wait six months to get them." Showing some kind of evidence that people will buy your product is probably the most important thing you can do when trying to get into a retailer as a young brand. You can point to good press as "social proof" (validation from someone other than your stakeholders that the product is awesome). When Heroclip was approaching REI, *Outside* magazine had selected it for their Top 5 Must-Have Gear list. You can bet that Heroclip dedicated an entire slide of their presentation to this brag-worthy fact. If you don't have the press, think of other creative ways to show that your product will resonate with the retailer's customers. One thing Heroclip did that seemed to impress a new retailer was to show that people who "liked" the product on Facebook were 145 times more likely to also "like" the retailer, compared with the general population.

The second thing you can do is use sales reps. Since retail buyers are busy people and want to see a whole bunch of products at a single meeting, let's give them a bunch of products to look at! The thing is, you don't have a bunch of products, so we are going to jump in with other brands through sales reps. Independent sales reps are individuals or agencies who go to retailers to sell products on behalf of companies. Typically, sales reps work on commission only on sales they generate (we have seen commission rates as low as 4 percent and as high as 15 percent). The downside of being part of a bunch of products being presented to a buyer is . . . that your product is one of a bunch of products. If your sales rep thinks your product has the potential to generate a large commission for them, they may pull your product from their bag of goodies first, but there is no guarantee of that. Oftentimes, sales reps "forget" smaller brands, for the simple fact that larger brands with larger product assortments generate more revenue and hence higher commission. The key is to keep your sales reps excited about your product with product news, marketing materials, and incentives so that you and your product are always at the front of their mind.

When you are fortunate enough to make contact with buyers, the most important thing is to assure them that *their* customers will buy your product. Do your research on the retailer, figure out who shops there, and be able to say in three sentences why those shoppers will immediately "get" your product and clamor to buy one. Prepare a short pitch—those three sentences that describe what your product does and what benefits the retailer's customers will get from it. It is critical that your pitch is tailored specifically for your retailer. You do not want to tell the buyer at The Container Store all the ways in which your product can help Harley bikers. Know your audience well, and do not waste their (and your) precious time by giving them irrelevant information.

allocated space in the store for your product, and now there is going to be a big, gaping hole in the wall or on the shelf because you couldn't deliver—and the buyer is going to be the one taking the flack for it. She does not want to be in that position!

We are huge fans of creating a tsunami with your product, and retailers, especially large ones, really have the potential to grow your sales in quantum leaps. But, as we've said before, it is a long, difficult process. Preparing pitches and presentations individually for each retailer is time- and energy-consuming. Even in the best-case scenarios, if your buyer and her team love your product, the actual purchase order (or PO, a document that makes their order official) might take months to come. And even when it lands in your inbox, it might get canceled (we know of at least a few companies that have almost gone under from large POs being canceled). Some retailers will not only want to nickel-and-dime you on pricing but also make you sign a slew of agreements that basically say you can be charged for violating any number of requirements. Even something as simple as putting a label on a carton can entail a three-page guide on how to do it exactly the way they want, with the threat of chargebacks (money you're charged when you violate a requirement). Retailers may also want generous payment terms that are up to ninety days—meaning that they will pay ninety days after receiving your invoice. (See the section on payment terms in Chapter 7: Financing Options, page 149.) Some retailers will want a guarantee that you will buy back the inventory if it doesn't sell. Some will decide to discount your product heavily to attract buyers, which will royally piss off other retailers (see MAP policy in Wholesale Terms, page 88). Sometimes you will sell to a retailer only to see your product end up on Amazon, priced ten cents below your listing, and essentially competing with you (and you won't even be able to track down the identity of the merchant because it is selling under a different name).

We don't mean to be Debbie Downers when talking about wholesale; we just want you to be prepared. We are all for you going after retailers and getting them to help propel your product to every corner of the world. You will get nos, but you will know that it is not you, just the nature of the beast. If you do get a no, don't just pack up your ball and go home. Find out why the answer is no, and see if you can help the buyer see things differently. One of the first retailers Mina got Heroclip into was a small six-store chain of gift shops called Fireworks (you can find the world's coolest things in Fireworks stores, and now there is one at the Seattle–Tacoma International Airport). She cold-emailed an address on the website and heard back from the "Chief Firecracker" herself who said no. With a couple more emails, Mina learned that she thought the MSRP (manufacturer's suggested retail price) on the product was too high compared to what she saw for items in the same category. So Mina sent her a short presentation showing competing products and all the additional benefits Heroclip offers at a slightly higher price point, and then brought out her crowdfunding and DTC data to show that there *was* already a proven market for the product at the MSRP. The Chief Firecracker was convinced and not only sent a purchase order but is a huge supporter to this day. Along the same lines of not giving up, if one buyer at a retailer says no, see if you can contact another one. Buyers are humans, after all, and have biases that may have nothing to do with your product.

A COUPLE MORE THINGS ABOUT SELLING

We've briefly covered the three main ways you can distribute your product, and at the risk of overwhelming you, we would like to mention two more distribution strategies that might come up.

The first is going through a distributor. To put it simply, you would sell to a distributor who then sells to a retailer who then sells to the customer. Yes, it is a long chain, but, theoretically, the advantage of using a distributor is that they would buy in even bigger quantities than a retailer since presumably they would be selling to many more retailers than you could directly. They warehouse the goods and then ship them as retailers place orders, eliminating the need for you to individually service each of these retailers (some of them may be very small).

The second term you might hear is *drop-shipping*. While an online retailer might love your product and want to offer it to your customers, they may not want to warehouse it or take the risk of buying it in bulk. In this case, they may want you to ship directly to their customers—in other words, they send you the orders they receive and you ship from your warehouse. The popular site Zulily.com is a good example of a drop-ship retailer. As a customer, you might pay Zulily for all those adorable baby shoes you buy, but when you receive them, you will see that they came directly from the seller.

Check out Wholesale Terms (page 88) to see some other phrases you might need to know in your quest to dominate the universe through retailers!

Comparing Margins

DIRECT TO CONSUMER

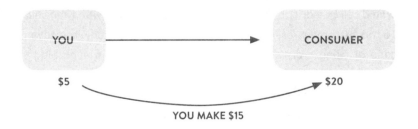

YOU

CONSUMER

$5

$20

YOU MAKE $15

THIRD-PARTY MARKETPLACE

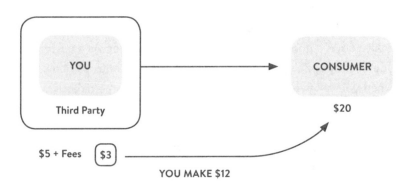

YOU

Third Party

CONSUMER

$20

$5 + Fees $3

YOU MAKE $12

Note that the $3 in fees above is just an example.

WHOLESALE TO RETAILER

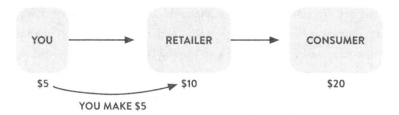

WHOLESALE TO DISTRIBUTOR

A Final Word

Margins are important! Oh, margins. If only InventHers didn't have to worry about them. Up until now, in this chapter we intentionally focused on explaining various distribution channels without referring to margins. However, it is critically important that we keep in mind that the different distribution channels come with different margin assumptions. A rule of thumb is the longer the distribution channel, the more markups there are and the lower your price to whomever you are selling. This is because whether someone buys your product on your website, on Amazon, or at Target, they are going to expect to pay about the same price. Working backward from this MSRP, if everyone has to make a bit of money, in order to keep the MSRP the same, you will have to take less—check out the math for the distribution strategies for a hypothetical product that costs five dollars to make, with a consumer price of twenty dollars, on page 84.

We bring up margins because along with all the pros and cons of each distribution channel we mentioned, your decisions are going to also be based on margin requirements, yours and whomever you are selling to. For some products and brands, using distributors will not even be a viable option because their margin requirements will basically leave you with nary a penny to show for all your hard work. For other products and brands, a low margin could be a very good trade-off if their pricing can bring massive volume. You decide what your business can handle and how you want to grow your business.

So, armed with all this information, it is time to start bringing customers to wherever you decided you want to sell.

Wholesale Terms

- **Chargeback:** What your reseller charges you for violating certain agreed-upon shipping procedures or quality standards.

- **Consignment:** A retailer pays you only after your product has sold.

- **Distributors:** Individuals or companies that stock your products and resell them to retailers. One reason retailers prefer to buy from distributors rather than individual brands is that they can use a single entity to buy an assortment of products from different brands. It's a one-stop shop.

- **Drop-shipping:** This happens when an order is taken from a retailer's site and the product is shipped directly from the manufacturer to the consumer. This means that the retailer never holds inventory of the particular product.

- **Group buying:** Since in most cases higher volume means lower cost, retailers may band together to form a buying group to increase the quantity they can buy. The purchased goods are then distributed to the members of the group to be sold independently.

- **MAP policy:** MAP stands for minimum advertised pricing. This policy is to prevent your resellers from discounting your products heavily in a public forum. You want to prevent this because heavy discounts that are advertised by one retailer could easily lead to heavy discounts from other retailers selling your product as they try to compete for business. This leads to a downward spiraling of your product's price, which is not good for anyone or anything, including the perceived value of your brand.

- **Reseller:** An entity that buys from a manufacturer and resells to a retailer or consumer.

- **Retailer:** An entity that sells to the consumer (i.e., a store).

- **Sales reps:** Folks who act as your company's representatives and pitch your product to retailers. Independent reps generally represent multiple brands and work on commission.

- **Sell-through:** How quickly your product sells.

Chez Brungraber
Gobi Gear

TAILOR-MADE FOR SCRAPPY GROWTH

Chez Brungraber knows what she wants, and always did. With her business, Gobi Gear, she wanted to grow it in a way that fit into *her* life and *her* budget, and not someone else's idea of how a business should grow. Unwilling to give up her day job or to take on the additional responsibility of having investors (or giving up equity), Chez has grown her business into a model of scrappy, controlled success that adapts to the changing circumstances of her life.

Chez's "aha" moment came when she realized for the millionth time that she didn't have time to be rummaging around her pack. As a wildlife biologist with a passion for botany, she spent a lot of time on the road, working in Kenya, Tanzania, Nepal, New Zealand, and other far-flung locations, but everything else needed to be carried on your back. After dumping out her backpack one too many times to find what she needed, she made a quick fix. With a sewing machine, she sewed a bag with sectioned compartments to keep all her gear organized. It was a hit with other trekkers on the Annapurna circuit, and she realized she had a marketable idea: Gobi Gear—compartmentalized, lightweight, packable bags to keep your pack tidy.

Chez launched Gobi Gear while continuing her other work, which was possible due to its seasonal nature. It took a year to nail down a factory, create a website, and start marketing. She found Twitter was a great channel for getting the word out—she was able to identify travel bloggers and Tweeters who were using certain hashtags. She sent out samples and they'd review her bags, and then offer unique discount codes to their followers. That way, Chez could know who was helping with sales and who was not. Then, another big step was a simple website pop-up for gathering email addresses with an offer of a discount. She was able to grow her list of customers to 30,000 and every email blast resulted in sales. More progress: she found some affiliate partners, like GearJunkie, that would promote her product for a cut of the sales.

Chez had an excellent plan for controlled growth that fit in with her lifestyle—seasonal work, coupled with running a start-up during slow months. She was able to find a path that worked for her. After she had children, she was able to step back and pay an employee to take over many of the tasks. Gobi Gear continues to grow, and now she is thinking of an exit. That is what success looks like for her.

"It's taken on life of its own now," says Brungraber. "It's okay to admit you're not the person to take it to the next level. I didn't have kids five years ago. I don't want to shortchange anyone, including my family. I will see it move to someone who has bandwidth to take the amazing places I know it's going to go."

Chapter 5: Marketing
Shouting from the Rooftops

Now you know your options for *where* customers will buy your product, but how will they learn about it in the first place? Unfortunately, Ralph Waldo Emerson's famous quote "Build a better mousetrap and the world will beat a path to your door" is no longer true (if it ever was). Even the most brilliant products will die a slow, painful death if nobody knows about them. This is where marketing comes in.

Simply put, marketing is how you get potential customers interested in your product. When introducing a new product, you'll have to put together a basic marketing plan—and you should revisit it at least quarterly. There is no point in creating a kick-ass product if no one knows about it.

Do not confuse marketing with advertising. Advertising is one small piece of a marketing plan. Marketing is all about strategy;

advertising is one way to execute that strategy. For example, say you have invented a cat collar with a speaker so you can talk to your cat while you're at work. (Think of it: your cat can now ignore you even while you're gone!) You might take out advertising in *Catster* magazine or buy digital ads on pet websites or Instagram ads for followers of #Caturday. That is just advertising. Your marketing strategy might aim to reach cat owners who earn over $100,000 per year and don't have children. Advertising would be a good start, but there's more. You might sponsor events at cat clubs, have a booth at cat competitions, have a partnership with a local rescue operation, team up with corporations to offer your collar as an employee benefit—you get the idea.

Marketing cannot be an afterthought. We know you're spending all your time on product development and refinement, fund-raising, dealing with manufacturing and distribution, and all the other myriad tasks that an InventHer must do. But even though you may think your product will sell itself, it won't. Putting together a marketing plan means sitting down and brainstorming.

The good thing is that during the course of your product and customer research, you already gained the basic ingredients to good marketing. The holy grail of marketing is what is known as the 4Ps—product, price, place, and promotion. We have already highlighted how important it is that your product, pricing, and place of sale align with the lifestyle, habits, and traits of your customers. Now we'll talk about the fourth P—promotion—in more detail, which involves coming up with a fantastic message about your product, then shouting it out where potential customers can hear you. Let's talk about your message and then take a look at marketing tools.

Four Ps of Marketing

Here are the 4Ps of marketing that you need to know: product, price, place, and promotion. These were first developed in the 1950s and have been added to over the years. But the original four remain the pillars of any marketing plan. Each element needs to be addressed, and together they form the foundation of a strategy for a bestselling product.

PRODUCT

The entire product is more than just your new invention. It includes the packaging, features, add-ons, warranty, return policy, customer service, and more. It's the whole enchilada—think of Apple products. It's not just that cool new MacBook Air that is the product; it's the experience you have when you buy it, unbox it, open the lid, and fire it up. It's the guy or gal with a headset who greets you at the Genius Bar when you come in with a tech issue. The name is part of the product. Apple has clearly nailed marketing on all levels, and it's why they are often emulated. Their product is more than just computers or phones.

PRICE

There is not just one price for a product. There are retail prices, wholesale prices, bundled prices, and discounted/rebated prices. Pricing can be a complex subject, but the basics are this: You need to make money off the product. A consumer who buys your product from your website, where you have to package and ship to them, will pay more than Target, who wants to buy 100,000 units. Want to have a holiday sale? Great idea! Make sure you set a price that can be reduced for periodic sales but still makes money.

PLACE

Where are consumers going to find your product? We addressed the basics of this in the previous chapter, but it can get a lot more granular. When you shop at the grocery store, think of the end of the aisles, the endcaps. That is the most coveted space for groceries, and products are rotated onto endcaps frequently, as manufacturers pay big money for that prime location. It's the same for any product: Where will it get the most eyeballs? Place can be physical, like a store, or digital, appearing on a highly visible web page. If you are selling that talking cat collar and a Google search of "electronic cat collar" doesn't list you as one of the top search results, that's a miss.

PROMOTION

Advertising falls under this P, as does your messaging: How are you telling the story of your product? Why should people buy it? How will you convey the message? How the message comes across can be as important as what the message is. Promotion includes public relations efforts. Hearing about your product from you is important, but hearing about it from television shows, news articles, bloggers, and social media influencers is even more powerful and will help your product fly off the shelves. Think about how to promote your product to your specific audience.

The New Ps

While the 4Ps remain the backbone of marketing, new ones have been added. (Of course, these critical marketing elements must begin with P, because . . . well, we don't make the rules. They just do.) They focus on business practices, and are good to keep in mind.

PROCESS

This one is about automation and scaling. Determining how your business can be more efficient is key to growth. It can be as basic as simplifying payment for orders, such as taking credit cards at events with an app such as Square that transforms your mobile device into a credit card machine, and as complicated as syncing business processes like inventory management and invoicing with a custom software integration. Look at every step of your business, especially the customer experience, and figure out how to make your operations fast and seamless, for customers, employees, and you.

PEOPLE

This P is all about the team. Maybe you've heard of the war for talent? Finding the right people is no easy feat, particularly for an emerging business. You might think that anyone with some good experience and a cheerful personality can get the job done. You'd be wrong. You will need extraordinary people to take your business to the next level. Some sage advice: Be slow to hire, fast to fire. Take the time to find the right mix, but don't hesitate to cut your losses when someone isn't working out. You can't afford to waste time as you grow. (See Chapter 8: Building Your Team, page 167, for more on this.)

Your Message

Any marketing professional will tell you that without a clear message about your brand and product, your marketing dollars are completely wasted. Creating a clear message is exceedingly (and surprisingly) difficult to do, especially when you have a product that can do a lot of things for a lot of people. Your message is the story of the product. There are many ways to develop a compelling and clear message, but here's how we like to think of it.

A message has several components, depending on the product and the market. But there is usually an emotional component (let's acknowledge once and for all that shopping is an emotional activity), a logical component (where the buyer thinks through the benefits of the product), and a technical component (the features of a product). After a customer is hooked by the emotional part, she moves to the logical and technical components to support and reconcile her emotional purchase decision, and then ultimately buy it. Take a look at the brand pyramid (at right) from one of our favorite branding experts for a hypothetical invention that whitens teeth. The value propositions of the "emotional" layer can be particularly difficult to pinpoint and articulate—Mina hadn't even realized why she had been questing for white teeth all her adult life our branding expert provided this example!

We know as well as anyone how difficult honing your message can be, and it is likely that your message will evolve and get tighter and more compelling as your company and product develop. Test your messages with friends, supporters, customers, and some small campaigns with strangers. Like everything else in bringing a product to market, the key in messaging is to learn from feedback and failures until you have something that resonates both with you as the InventHer and with your customers.

The Hypothetical Product

The Revolutionary Never-Fail Teeth Whitening System, $150

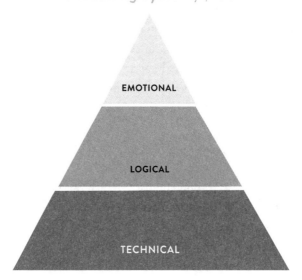

Emotional: *Purchase Decision Made Here*

- My ex-boyfriend will be so jealous of a new, perfect smile.
- People used to compliment my teeth when I was younger; I miss that.
- Even though I'm a parent now, I can still look vibrant and glamorous.

Logical: *Purchase Decision Support*

- $150 is nothing if my new confidence gets me that job promotion.
- $150 is worth letting my friends see me smile again.
- $150 to look ten years younger? That's only $15 a year or 4 cents per day!

Technical: *Further Purchase Decision Support*

- They use the best ingredients in their whitening system.
- Their product comes with a lifetime guarantee.
- Their proprietary whitener already has a 4.5 out of 5 star online rating.

Courtesy of Greg Leschisin, Creative Director / Principal, Hone & Woo

The remainder of this chapter is about the various tools you can use to broadcast your message. Just like it is virtually impossible for a small new business to excel at distributing through every available channel, it is likely beyond your resources (of energy, time, money, and sanity) to use every marketing tool available. Pick and choose based on the particularities of your product and your resources. The big question to ask yourself: How and where do I tell the story of my product to the most people who might be interested?

Public Relations

Getting to the public through media can be a full-time job for many. Most big businesses have dedicated PR departments or outside agencies who manage this. They build relationships with media outlets over time, and pitch them stories and breaking news about their clients. Until you get big and hire someone to handle this, you can do some basic PR outreach for yourself. It can be as simple as emailing the local newspaper or radio station about your business. Make sure you offer a real pitch—why should they care about you? Maybe the local newspaper's business editor would like to hear "Local woman invents shoes made from old tires, keeping them out of landfills." Other relevant news might be moving into an office space, hiring employees, winning an award, or getting funding rounds. Or you might just see if they are interested in doing a profile on you and the story of how you invented your product. Now is not the time to be modest—you've got to toot your own horn. Believe it or not, newspapers, magazines, websites, and local news programs are all constantly looking for new stories that might be interesting to their readers, viewers, or listeners. Everyone loves a good origin story:

the homemaker turned inventor who is trying to reimagine the pet collar or travel pillow, or who is building a better mousetrap. Many a product has leapt to stardom through selective PR (read more on page 120 about the inventor of the Squatty Potty, whose turn on *The Dr. Oz Show* was a game changer for her).

Don't think you need to be in the *New York Times* or *Oprah* magazine for effective PR. Local media outlets work great. Think outside the box for media too . . . popular blogs are great ways to reach new audiences. Contact relevant bloggers and see if they are interested in hearing and sharing your story. Podcasts are another outlet—think about who might be interested in having you on their show to talk about your product. A podcast or blog that is focused on green or sustainable living practices might want to hear about your clothing made from recycled soda bottles. The key to PR is outreach. You have to go out and find people with big megaphones to shout your story. Good PR builds awareness of your brand, and brand awareness leads to sales.

Your Own Blog Posts

A blog on your website is a good place to highlight some unique aspects of your brand. You might want to share some behind-the-scenes looks at the company, profile team members or customers, do Q and As with relevant people, or talk about the uses of your product. Mina has used her blog to showcase ways to use the Heroclip that people might not have thought of, including terrific customer uses. She uses social media channels for this as well, but a blog allows for a longer-form exploration of a subject. It also can highlight Mina and her team—whether discussing the pros and cons of crowdfunding, describing how the team went about testing the Heroclip, or explaining the process of making new iterations or designs, a blog really helps a brand (and the people behind it) tell its story. Best of all, it's free!

Guest Blog Posts

Once you have a blog, it's a never-ending search for relevant, interesting content. Seems easy—those first few weeks or months, you'll have many new things to talk about. Then comes the dry spell . . . what to talk about this week? An inventory of the company fridge? A profile of the FedEx delivery guy? Nope. For a marketing boost, get some guest bloggers, preferably ones with some influence. Influence can come in many forms, but number of social media followers is one good measure. If you've got a new exercise product, reach out to any of the many fitness bloggers and see if they'd like to try it out and blog about it. Or just have them write a guest blog for you on a non-promotional topic, like top ten fitness resolutions for the New Year. If you've got a pet product, tap someone who runs a dog-walking service or doggie day care. Ideally they will share this blog post with their audience—whether on their social channels, in a newsletter, or on their website. You've just introduced their network to your website and, by extension, to your product. Hilary is frequently asked to guest blog for various start-ups or companies. As a travel writer for several publications, she was recently asked to guest blog about travel tips for a friend's start-up that sells concierge-type services that include vacation planning. Then she shared this post on her LinkedIn, Facebook, and Twitter accounts. She got questions about the service from friends and connections, and could then point them back to the website, plus give a little personal endorsement for the service.

These interactions can be more useful than a shotgun approach to strangers, as connections in your network are more likely to consider something from a trusted source rather than an ad. Don't be afraid to reach out and ask people you don't know to blog for you—most people like to talk about their expertise or opinions. And again, the blog doesn't need to be promotional. You're getting relevant content for your site—and establishing your website as a trusted source of information. A blog post from a dog walker on dog-park etiquette is super useful for a website selling customized dog collars. When that dog walker shares on their personal Facebook page, "Hey, I wrote a post on dog-park etiquette for SnazzyCollars .com—check it out!" they are bringing eyeballs to your blog, and by extension to your website. After reading about how it is not cool for dog owners to bring aggressive dogs to a dog park, they might just click around the site and be induced to purchase a new collar for Bark Twain or Woofilicious.

Sometimes it's even worth paying a guest blogger. There is no set rate that is considered appropriate. Friends, family, and colleagues will often do it for free. But if you've got the perfect influencer who is willing to blog for a price, you've got to weigh several factors: their reach (how many eyeballs can they get on your product?), their audience engagement (their audience reads, but do they buy?—ask for stats), the quality of their posts (will it include video?; is the writing snappy and shareable?), and your budget. With all of marketing, it's about figuring out the biggest bang for your buck. The $500 they want might be better spent on a Facebook ad. Do your research but also trust your gut.

Events

Time to get off-line and into the real world. There is nothing like face-to-face interaction to showcase your product. You don't have to start in a big retail store. Test the waters with smaller venues. The best thing about in-person sales is the interaction with and immediate feedback from customers. Sell, but also listen. Hear what folks are saying. Some places to think about (depending on your product) are farmers' markets, craft fairs, and church bazaars. Crafty things like bags made from upcycled materials will be a good fit for a farmers' market or craft fair. It is often a minimal cost to rent a booth or table space at one of these, and you've got a ready-made audience coming in to buy. You'll get to interact one-on-one and ask questions. One bike-apparel company we know got started by buying a booth/exhibition space at the Seattle Bike and Outdoor Show, a big expo in early spring. They sold out of their product over the two days of the show, but also caught the eye of a local bike shop that was interested in carrying their product. They were launched.

Another on-the-cheap marketing opportunity is school or non-profit auctions. These are big fund-raisers, and organizations are always looking for donations to put in a silent or live-bidding auction. Find an event that matches your audience and offer up a sample. Depending on how big the list of invitees is, you could get in front of a whole new audience. Usually your item will have a detailed description either posted in front of the display or in a catalog at a silent auction. It will definitely get a look from almost everyone attending, as that is the point of an auction. At one recent elementary school auction, the hottest item was a basket with some cannabis-related products from a local company. While only one lucky winner went home with it, the dozens of people who unsuccessfully bid on it, or just viewed it, left knowing the name of that business and were free to go home and purchase more products. (Note: The next most popular item was a vasectomy package from a local doctor—these two items might give insight into the challenges of modern parenting!) Bottom line: don't miss an opportunity to get your name out there for very low cost.

Giveaways

Events, auctions, and guest bloggers are really just a subset of one of the strongest marketing tools: the giveaway. While not free, it's usually far cheaper to give away product than to pay for traditional advertising. Everyone likes free stuff—you can give it to an Instagram influencer with the hopes they will post a photo out of the goodness of their heart. You can give product to bloggers for them to keep and use, or to gift to their own followers. You can also do a straight contest give-away on social media. A common practice is to have people sign up for an email list or follow your social channel to be eligible for a drawing. Ideally you want to get something back in return for a giveaway (duh), and the chance to email them later with special offers or for them to see a future social media post is what you're getting.

Brand Ambassadors

Brand ambassadors take the idea of influencers to the next level. Brand ambassadors are basically public users of your product who actively evangelize for you. They can be celebrities or just superen-gaged fans who do it for some swag or discounts. At the top of the pecking order are the celebrities—Kylie Jenner was the highest-paid

Instagram brand ambassador in 2018; she reportedly earns $1 million for every post in which she hawks shoes, weight-loss products, or similar items. Soccer star Cristiano Ronaldo makes more than $750,000 per post. Why do brands pay this much? Because it works. It's all about the eyeballs, *engaged* eyeballs. Kylie Jenner has more than 123 *million* followers, all of whom just saw her post her brand-new shoes. And the people who saw it really like her, and some want to emulate her. Those eyeballs are more valuable than 123 million people watching the Super Bowl or seeing a billboard in Times Square. They are the perfect audience, and they open their wallets.

Thankfully, you don't have to spend a million dollars or hire a Kardashian to get yourself a brand ambassador. You can recruit customers, friends, and colleagues. Customers who love your product are a great start. You can launch a customer brand-ambassador program pretty easily. Create a simple application page that includes some parameters, such as whether your customers are fun loving, outdoorsy, pet loving, busy parents, avid travelers—whatever fits your brand. Then offer some perks—discounts on product, or even free product—in exchange for sharing on social media or in another medium. Hilary set up a program like this for a client with a music-oriented product. The form asked people to list their social media channels and number of followers. Then each person who met the criteria was given a customized discount code to share with friends and followers. Hilary's client could track how many sales came through each ambassador. Five sales resulted in a 50 percent off coupon for another purchase, ten sales was a free product, twenty was a bigger prize . . . you get the idea. The key is to come up with a program that motivates a brand ambassador to go to work for you.

Affiliates

To truly create a tsunami in selling your product, you have to be everywhere. But you can't. Affiliates are folks, generally online, who will write about your product in the context of their own website, promote this content, and get a cut of whatever you sell. Many participants in affiliate programs are small bloggers, but this is where numbers come in. Imagine having 100 affiliates with 1,000 followers each—that is a reach of 100,000! Generally, because the affiliates create their own content, there is minimal work required of you beyond giving them assets they can use. There are even platforms that provide unique links for affiliates so that sales can be properly credited to them and commissions automatically calculated and paid out. You can offer affiliates whatever percentage you want, but, needless to say, the higher it is the more motivated they will be.

Customer Reviews

Ideally, your best brand advocates are those customers who love your product and evangelize it to their peers without prompting or perks. Word of mouth is still a powerful tool, even in the digital age. But people also listen to strangers. Think of Yelp. People can be swayed by one really good (or bad) review of a restaurant or hotel. Lots more will be swayed by *many* good (or bad) reviews of a place. There are loads of product review sites, and it never hurts to ask a happy customer if she'd mind writing a review or rating your service on one of these sites. Google's review site has been gaining popularity, and there are niche review sites for everything from electronics to toys.

Find out where your product might be reviewed, and reach out to happy customers and your network to post some positive reviews.

Cross-Promotion

Cross-promotion is a fancy way of saying, "You scratch my back; I'll scratch yours." It's a well-accepted practice on crowdfunding platforms such as Kickstarter and Indiegogo. On those platforms, when you start a campaign, you will see other campaigns with overlapping audiences. You can partner with one of them and introduce your backers to the their campaign while they introduce their backers to yours. But make sure it is a good fit. If one campaign is trying to fund a new photography drone and the other is making bamboo tableware, there's probably not a ton of crossover. But if one campaign is pitching a new TSA-friendly backpack for travelers and the other is promoting a new travel neck pillow with speakers—this might be the start of a beautiful friendship. You are both appealing to the same audience, but not competing with each other, so why not boost the marketing reach of both your products? That is the heart of cross-promotion.

Cross-promotion isn't limited to crowdfunding. Featuring guest blogs from other products or services on your site is also a form of cross-promotion if they get a plug in for themselves, and/or if you link back to their web page. Think of who shares an audience. That dog-walking service might want to share mailing lists with the new SnazzyCollar folks—the dog walkers might offer 15 percent off a new SnazzyCollar with a special code, while the SnazzyCollar customers might be introduced to the dog-walking service through a free dog walk. At an event, like a trade show or farmers' market, keep a stack of coupons or handouts for your partner while they do the same. If the cost of a booth or exhibition space is too high for you individually, share one. Another great way to cross-promote is through cosponsored contests: You can always give out your own things, but when you pair them with other merchandise, you're amplifying the value and potential interest. Give your product to cross-promotional partners to give away in their contests, while you get some of theirs to do the same. Share the contact information for entrants. How about a cobranded advertisement in local papers or online? The bottom line is figuring out a partner with a shared audience, shared goals, and a similar enough vibe that it makes sense to throw business each other's way.

Cross-promotions don't always have to be between two product sellers. Nonprofits or charities can also make great partners. Many nonprofits partner with brands to get a portion of proceeds donated to their charity. Often they will have a reach that you can only dream about. Approach the local Humane Society about a partnership for your pet product—a dollar from every sale goes to the Humane Society! Or create a special discount code for them. What do you get? Their dedicated audience of thousands of pet owners who support the charity. Many will need a pet product—they might as well buy one that supports a cause dear to them. (Think of the countless pinkified products for breast cancer research.) The charities don't have to be large—it could be your local Girl Scout troop, the high school marching band, or a neighborhood conservation group. Do some good while doing good for your product—cross-promotion benefits you both.

The absolute pinnacle of cross-promotion is selling each other's goods on your own sites. This requires a deeper partnership. You can even offer "special editions" of your product for your partner.

Subaru offered an L. L. Bean edition of their car, a great example of serious cross-promotion.

What we've talked about so far is usually categorized as traditional marketing. But in the digital age, there are some specific marketing channels that need to be considered as well. Of course, all marketing overlaps, but the following are critical for your e-commerce.

SEO (Search Engine Optimization)

This is something to keep in mind as you develop your website. SEO, or search engine optimization, is about how easy it is for users to find you in a search. There are multiple search engines, but Google is the one most people use. You want your site to be as search friendly as possible so that people who don't know you exist will find you. It's all about traffic. There are digital professionals who devote all their time to making sure Google finds their clients' websites, but the key thing is to have multiple keywords and tags that Google can identify as it crawls, or reads, the website. If you're selling a digital pet collar that can track your pet, you might need to have multiple phrases, such as *cat collar, dog collar, electronic collar, digital collar, tracking pet collar,* etc. Think of all the phrases that people who are looking to buy might type into the search field.

Besides the correct words, Google will be looking for many other things, all part of a complex algorithm that smarter people than us try to figure out daily. But some things are key: your site has to load fast. Period. No one wants to wait for loading, and Google knows that and will not reward you with traffic if your site is slow. They will also look to see how many other sites link to you—another sign of credibility and quality that will raise your ranking. It's another reason things like partnerships or press attention help—these are other sites that will provide a link to yours.

Google Ads

To get traffic, you need rankings. Let's say you want to rent a car in Houston. If you Google *rental car in Houston*, what comes up first? Most people are not going to scroll to the fourth page of results. The car rental agencies that come up first, second, and third are going to get the lion's share of business. As a result, companies jockey to make sure they end up on top through Google Ads (formerly Google Ad Words). This is the heart of the Google business model—they auction off the best keywords for top dollar. However, it's a bit more complicated than that. They use an algorithm that also factors in the quality of your site—they want their users to have a good experience, so a shoddy site of low relevance to the keywords up for auction will rank below a high-quality site that bids less. Confused? Don't worry—keeping track of this stuff is a full-time job. What's important is knowing the best keywords on which you can bid. Don't waste time or money with really broad keywords or not-quite-right keywords. There might be some trial and error. The first step is creating a Google Ads account at Ads.Google.com. Setting up the account is fairly straightforward. You just need a website and a credit card.

Social Media Platforms

It may be tempting to sign up for all the social media channels. But don't overextend yourself, especially if you're a small operation. Social media posting for a business is a commitment, and it takes time and, depending on your strategy, money. Assess what channels are best for you and which ones you feel most comfortable with, and focus your efforts there.

 HOT TIP: It's better to do one or two social media channels really well than to do five haphazardly.

FACEBOOK

Let's start with the low-hanging fruit. If you haven't set up a business Facebook page, run to your computer and do it. Facebook is the 800-pound gorilla when it comes to marketing. First, the basics. Set up a page. Use engaging photos and simple messaging. Share it on your personal Facebook page, and ask your friends and family to "like" it. Your personal network of friends is your first group of customers. Many a successful business has started with word of mouth through friends and family. Today, that often means Facebook. It has an amazing reach—more than 2 *billion* active monthly users. Surely some of those folks want to buy your product! Start building an organic (unpaid) presence, then think about a small budget for paid advertising to take advantage of Facebook's enormous reach.

You might have heard in the news that Facebook has been dealing with some privacy concerns—they have a ton of data on their users. Bad for privacy, great for commerce. You can do highly specific targeting to reach the very audience of buyers you need. Looking for college-educated, forty-to-forty-five-year-old moms of small children who are politically liberal, who enjoy surfing and horseback riding, whose household income is in the top 10 percent of all zip codes, and who live in Cleveland? Yeah, Facebook can target exactly those people to sell them your kids' surfboard with horse graphics that benefits the Cleveland Democratic Party. (Warning: this seems like a niche product.)

Like all social media, Facebook provides an excellent opportunity to interact directly with your customers and potential customers. You can ask and answer questions—key to your market research—and see customer-service issues before they become problems. We're big fans of Facebook Live—the feature that allows you to broadcast live from your computer. Think of the possibilities! You can demo how your product works, give a behind-the-scenes look at your work space, interview team members or other guests—anything! Mina and her team do these regularly, and they get input from the community on how people are using their Heroclips. The real-time aspect is compelling—while high-quality recorded video will always be key, there is no match for live authenticity and realness. Facebook Live is also a great way to make big announcements, like the launch of a new product or promotion.

TWITTER

Twitter is a love-it-or-hate-it platform for marketing. While it doesn't have the power that the Facebook juggernaut does, it still has some great uses. Twitter really shines around events—there are hashtags for many events, from the Super Bowl to local trade shows. There are more than 313 million monthly users, and you can capture some of these eyeballs without spending a dime.

Hashtags are a way of sifting through the noise of Twitter. You might follow certain ones like #Seahawks if you're a Seattle football fan, or #goldendoodle if you're a fan of that dog breed. Think of your audience: Who are you trying to reach? Now find where they are on Twitter. For example, if you've invented a bike-related product, you'll want to go where the cyclists are. Some quick work with Google shows that many top bike-related hashtags are more specific than just #bike or #cycling or #riding. True fans are using #baaw (bike against a wall, for shots of their bikes propped against walls), and #foreverbuttphotos for shots of their teammates or cycling buddies riding just ahead of them or #kitspiration for shots of cool cycling kits (outfits). There are groups using #BikeNYC and events are tagged #BikeExpo. Think of how you could showcase your product using hashtags that already get attention. If you're selling a new kind of bike shorts, get some product shots in action and tag them #foreverbuttphotos or #kitspiration. Or if you're going to be at an event, post a tweet that says, "Come visit us at #BikeExpo and enter to win one of our new designs!" Unlike Facebook, potential consumers don't have to be "following" you to see your tweets. They just need to be watching certain hashtags.

Of course, like all channels, you can greatly increase your visibility by spending money. You can sponsor tweets to show up in the timelines of people you want to target. You can target customers much like you can in Facebook, although the categories are different.

Twitter really puts the "social" in social media. Of all the channels, it's the easiest when it comes to having direct dialogue with customers and prospects. You can run a poll, ask direct questions, get feedback, and solve customer-service issues. There is nothing as immediate as Twitter for interaction.

HOT TIP: Sponsor tweets to appear in the timelines of your competitor's followers. If you're selling those cute bike shorts, target the followers of another popular bike-short brand, cycling race, or magazine. Your audience is right there!

INSTAGRAM

Instagram is a photo-based social network which is now owned by Facebook, so you can run ads for both directly from the Facebook platform with the same granular targeting. It might be even better for impulse buying than Facebook for certain items. With more than 1 billion users and growing, it is answering the call for e-commerce. The changes are happening faster than we can explain here, but "shoppable posts" are now making it easier than ever to buy straight from your phone. (The days of being limited to having only one link in the bio are over.) Products that are photo friendly (duh) can do really well here. Apparel does really well (who doesn't want those cute boots that just popped up in their feed?). Instagram stories (that only last for 24 hours, similar to Snapchat stories) are another way brands are showcasing products. The ability to use video here is also very important—there is no better way to show how a product works. Mina can show a little video of how the Heroclip opens and swivels. Hilary may or may not have just bought a new purse from Instagram because the video showed the inside with many different pockets and zippered pouches that looked useful. Impulse buying is a thing, and Instagram is here for it.

LINKEDIN

Long known as *the* professional-networker platform, LinkedIn has grown from being a résumé repository to a full content platform. It's not a commerce platform, but a growing business, especially one looking down the road to fund-raising, should consider building a business presence on LinkedIn. It also lets your employees have a place to list their employer. It gives potential investors a quick snapshot of your business: where it is located, how many employees, and so on. Relevant news can be shared here as well. Got a new round of funding? Share this on LinkedIn.

SNAPCHAT

Snapchat is the current darling of the social media scene, particularly within the millennial and Generation Z set. According to the Pew Research Center, 78 percent of US eighteen-to-twenty-four-year-olds are on Snapchat.[11] According to Mashable, 77 percent of college students use Snapchat at least once a day.[12] Do the math: Are you trying to sell to this demographic? Then maybe your business needs to be on Snapchat. This is a very tech-savvy audience—you better be ready to keep up the quality content and hilarious videos. Snapchat's ads can target your buyers. Geofilters can limit ads to a certain location for a certain time period.

PINTEREST

The digital scrapbook site has a dedicated base of users—many of whom are happy to open their wallets. It's an older demographic than the Snapchat crowd. Seventy percent of Pinterest users discover new products through Pinterest, and 66 percent buy something after seeing a brand's pins.[13] The majority of users are women. When choosing a social media network to pursue, don't ignore this one if it matches your demographic.

A Final Word

We know this chapter has a lot of information—don't get overwhelmed! We hope that the myriad options for marketing are comforting, because if one thing doesn't work, there are many other avenues to explore. Often we find that after a few trial attempts, a solid path to marketing will become clearer.

The last thing we want to leave you with is this: Always be talking up your brand. Anywhere. Anytime. You're the captain of the cheerleading squad for your products. You are the authentic voice for your brand and why it exists. Don't be shy about sharing your story. You never know who or what is going to catapult your product to stardom. As an InventHer, you must ABM: always be marketing!

Key Terms and Metrics

- **Ad spend:** Amount of money you pay a platform (such as Facebook) to reach the audience you want.

- **AOV (average order value):** The average amount of money people spend at your online store.

- **Attribution:** Figuring out what led to the sale. This can be notoriously difficult. A customer could see your Facebook ad today then go to your website and not purchase, but then two days later, they could remember the product, go on Amazon, and then buy it. Facebook would say that this ad wasn't successful for this customer, as she didn't purchase on the location the ad sent her to, but is that truly accurate?

- **CAC (customer acquisition cost):** Amount of money you spend to convert a potential customer to an actual customer.

- **CLV (customer lifetime value):** How much money a customer will spend with your business as a repeating customer.

- **Conversion rate:** Percentage of people who see what you posted online and then proceed to do what you ask.

- **CPC (cost per click):** Amount of money it costs you to get someone to click on whatever you want them to click on. This is determined by looking at the aggregate of money spent and clicks made (e.g., if you spend $1,000 on ads and 100 people click, that would be a CPC of $10).

- **CPI (cost per impression):** Amount of money it costs you to get someone to see your content. This is determined by looking at the aggregate of money spent and total views (e.g., if you spend $1,000 on ads and 5,000 people see it, that would be a CPI of $0.20).

- **CTA (call to action):** What you want people who see your content to do (e.g., buy your product, check out your newest blog post, join your email list).

- **Number of impressions:** Number of people who see whatever you posted online.

- **ROAS (return on ad spend):** Sales you make through an ad relative to the amount of money you spend on reaching people with the ad. ROAS is the ad revenue divided by the ad spend. ROAS of five would mean for every dollar spent on advertising, you make five dollars.

- **ROI (return on investment):** Profit you make relative to costs of a campaign (including labor, ad spend, cost of goods sold, and so forth). ROI is the revenue minus costs, divided by costs. This is expressed as a percentage. An ROI of 50 percent means that for every dollar you spend, you make fifty cents in profit. Note that what is included as costs in defining ROI is a gray area and may vary from company to company.

- **Website traffic:** Number of unique visitors to your website in a given amount of time (usually per month).

Judy Edwards
Squatty Potty

THE POWER OF PR AND POOPING UNICORNS

Judy Edwards was in her sixties when she faced a common problem: constipation. A medical professional advised her to use a stool to raise her knees while on the toilet. It worked, and she started trying to make adjustments to get the stool just right. Soon she had her son and husband working on prototypes, and the Squatty Potty was born.

The first Squatty Potties were wooden and painted in Judy's garage. Judy gave them out to friends and family for Christmas, with a handout they made about the benefits of pooping in a squatting position with information downloaded from the internet. The gifts were received with laughs and good cheer. About two weeks later the calls started to come in, with the recipients raving about how the product changed their lives. The Edwards family knew they had a hit.

They struggled through their first manufacturing challenge, finding a factory in China, but it was a slow process, as language and cultural barriers kept them from thoroughly explaining the concept to their representative. Eventually they found a native speaker who could explain the squatting concept, and a design was finalized. In a leap of faith, they ordered 20,000 Squatty Potties, which arrived in a rented warehouse in their Utah town. "How in the world are we going to sell these?" Judy thought. A good day saw ten orders come in. But Judy's son had some marketing savvy—he reached out to health bloggers, offering free Squatty Potties for review. They took him up on it, and blog posts began appearing, singing the praises of the little stool. Orders began rolling in.

Then came the first turning point: a call from *The Dr. Oz Show*. Judy's husband, Bill, took the call and assumed it was a prank played by his son. He laughed and said, "Sure, and I'm Barack Obama." He quickly realized his error. The Squatty Potty was featured on the show. Orders were now pouring in. Soon, another spike in sales hit. They didn't know why at first, but discovered that the product was discussed on *The Howard Stern Show*, with sidekick Robin Quivers gushing about it. Sales went through the roof. Other celebrity raves followed, including ones from Sally Field, Bryan Cranston, and Jimmy Kimmel. Apparently everyone really does poop and can use a little help. The Squatty Potty was a bathroom star.

Then came the second turning point: a call from the show *Shark Tank* and an invitation to pitch it on television, in front of an audience of nearly 5 million people. Within twenty-four hours of the show airing, they had sold $1 million worth of Squatty Potties. At this point, Judy and her family had spent almost no money on marketing—some fortuitous PR and high-profile spots had gotten the word out.

Judy and family were not content to sit back and coast. Now they had some money to work with, so they tackled marketing head-on. Their main challenge was how to talk about pooping without sounding gross. They turned to a small marketing agency recommended by a cousin. The creative team pitched several ideas, but were adamant that the best one was a video about a unicorn that pooped rainbow ice cream with the help of the Squatty Potty. Talk about a leap of faith! They took the risk—check out the results for yourself. Google *Squatty Potty pooping unicorn video* and watch it. We'll wait.

Amazing, right? The result was 66 *million* views in the first four months. According to Judy, they reached 1.5 million Facebook shares, a 600 percent increase in online sales, and a 400 percent increase in retail sales. This is the definition of a viral video. But while hilarious, its real value is the education it brings to the consumer. It found a way to talk about a taboo subject that wasn't, uh, crappy.

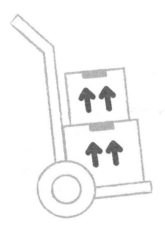

Lisa Fetterman
Nomiku

COOKING UP CUSTOMER FEEDBACK

Lisa Fetterman was in love . . . with the sous vide machine. She was introduced to its marvelous water-circulating cooking abilities when working in some of San Francisco's best restaurants. She began saving up enough money to buy one of the professional tools for herself, a $5,000 investment. Then, a first date changed her life in more ways than one. Her date, an expert in plasma physics, had an idea. "Why don't we go make our own sous vide machine?" They went to a hardware store to buy parts and soon had a DIY water-immersion circulating device for a tiny fraction of the cost.

Things really started to heat up. Lisa and the date, Abe, got engaged. They started teaching classes on how to build the machine yourself, then put together a DIY kit. Then they launched a Kickstarter campaign to sell the first-ever home sous vide device: the Nomiku sous vide. They made $500,000 in the first thirty days. A second Kickstarter offered an updated version with Wi-Fi capabilities and netted $750,000. Lisa and Abe moved to China for two years to learn everything about manufacturing. They wrote two cookbooks.

Shark Tank called. After an appearance on the show, sales increased tenfold in one day. Copycat devices popped up like weeds, but they pressed on.

Lisa and her husband could have declared success at any point—sales of the Nomiku were humming along. But the entrepreneurial spirit that drove Lisa to build her own device also drove her to do better. She read about Toyota's "go and see" principle—to see how something is really used, you must go to where it is used. She wanted to know how she could get people to use the Nomiku more. She reached out to customers who were in the San Francisco area and asked if she could come see them cook in their kitchens. She would observe and learn, and after every session she'd cook the family a meal. At last count, she has been to more than three hundred homes.

"They always said, 'We wish you could just cook for us!'" said Fetterman. "I listened. We decided to move into meals. Now we have meal kits. The Nomiku is almost like a Keurig for meals." With the new system, ingredients in the kits have RFID (radio-frequency identification) tags: wave them in front of the Nomiku and it will automatically set the water temperature and time for the food to cook. Just drop the food into the water, and dinner will be ready at the appropriate time.

Lisa and her husband picked a time-intensive but successful method of marketing, based on making their core customers happy and building a thriving referral system. They put a lot of effort into keeping these customers satisfied and continuing to serve their needs. For a niche product like the Nomiku and now Nomiku Meals, this strategy is paying off. The customers are bringing in their friends and evangelizing the brand to their communities. By listening to her customers, Lisa has opened up a new revenue opportunity in meal kits, a $3.1 billion market in 2018.[14] Nomiku is on track for more delicious success.

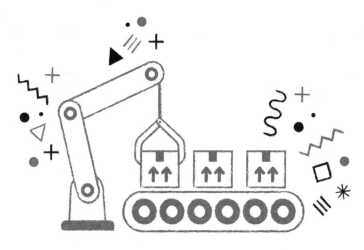

Chapter 6: Manufacturing

Getting It Made

Ideas are the first step to being an InventHer, but now you have to figure out how you are going to make your product at a scale you want. If you would like to turn your product into a growth-oriented business, it is absolutely critical that you find a manufacturing solution that is scalable, or able to accommodate the kind of growth you envision. Our InventHer stories include the various factors that go into how quickly and how large a business can and will grow, and these factors can be something personal, something situational, or something that is beyond your control. Mina knew she wanted her product to be a launchpad for a fast-growing company, and she learned the hard way the importance of having a great manufacturing partner and a reliable manufacturing process. We want to make sure you don't go through what she experienced along the way.

(Goods held hostage! A lawsuit! Delayed shipments! Unhappy customers! Canceled orders from her biggest wholesale account! Ugh!)

Manufacturing can mean a lot of things. It can be as simple as hiring your mom and a bunch of her retired friends to come to your house every day and sew your aprons with the built-in oven mitts. Or it can mean outsourcing all of your production to a facility in the United States or overseas.

One of the first things to determine is whether you will be using a domestic or overseas manufacturing facility. The common belief seems to be that companies choose to manufacture overseas because it is cheaper and they want to make more money. It is true that manufacturing overseas does tend to be cheaper if you look at just the unit price of a product. But there are other considerations that come into play when you start exploring factories. In this chapter, we'll go over the questions you should ask a facility (and yourself) when you are assessing and evaluating manufacturing options. Note that some manufacturers will be able to give you (*very*) ballpark estimates based on detailed sketches and homemade prototypes, but the majority will not be willing to seriously entertain making your product without a CAD (computer-aided design) file. Since manufacturing can be technical, in the course of guiding you through the manufacturing process, we'll also highlight useful terms you can learn and whip out to sound like an expert when you are interviewing potential manufacturers.

 HOT TIP: Make sure you know the jargon so potential manufacturers don't try to snow you.

What to Learn About Your Manufacturer

Minimum order quantity (MOQ): MOQ will come up a *lot* when you are exploring manufacturers. It simply means the minimum quantity the customer (you) is required to order for the manufacturer to work with you. By the time you get to the point of thinking about manufacturing, you will have some sense of what kind of demand your product may generate (from all that research you did) and some sense of what you can afford as your first order. Of all the things you need to find out about your manufacturer, you will want to ask about the MOQ first, as it will eliminate about 90 percent of factories for newbies who want to mass-produce their inventions.

When Mina first started thinking about manufacturing, her initial idea was that she would manufacture 500 units and see how they sold (this was before she decided to do a crowdfunding campaign). Being a former academic with a lot of quantitative analysis under her belt, she felt that 500 was the perfect number to give her enough reliable sales data and also wouldn't totally deplete her savings. The only problem? Every manufacturer she spoke with laughed in her face. Actually, they didn't laugh—they either said no or gave a ballpark figure that was obviously designed to make her go away.

What is the right MOQ? Well, that depends on what you're making and how you are making it. When Mina was touring factories, she visited one that had an MOQ of one, but it was a factory that made super-high-precision replacement parts for airplanes. In general, for Heroclip the MOQ seemed to be north of 5,000 units per order and 2,000 units per color. The MOQ the manufacturer gives you is not just about what it is willing to do but also the constraints it is working with. For example, to make soft-good products, a manufacturer will need to buy fabric from its fabric supplier, and fabric is sold in rolls of

1,000 yards. It would want an order that uses up all of this fabric (or at least pays for all of it). The MOQ will also depend on the production method used (which, in turn, depends on the desired quantity). For example, die-casting requires a lot of tooling and mold creation and requires a larger quantity than, say, programming a CNC (computer numeric control) machine to cut a similar product from solid metal. A good manufacturer who really wants to work with you will brainstorm together on production methods and make suggestions on ways to reduce MOQ and costs.

Questions to ask:

- What is your MOQ?

- What production method do you recommend?

Unit price, BOM, and payment terms: Once you have a few manufacturers willing to work with you, you will want to talk pricing. The price that they give you, the unit price, is made up of parts and labor cost. It is very important that you get a bill of materials (or BOM, as industry insiders call it). A BOM is basically a list of raw materials, parts, and labor required to make one complete product. Depending on what you are making, you will want to be as specific as possible and document the type of metal if metal is being used (did you know that there are dozens of varieties of aluminum?) or even the type and brand of nylon that will be used if your product is plastic. In the experience of our InventHers, many manufacturers are reluctant to provide a detailed BOM (actually, it's like pulling teeth). This is because aside from the work required to put a BOM together, which can be substantial, and the fact that a BOM binds a manufacturer to the use of certain materials and pricing, a BOM enables you to do an apples-to-apples comparison of costs when shopping around for a factory. So make sure to get one!

Another thing to consider when you are determining the unit price of your product is packaging. As you already know from being a consumer, there is a huge range in packaging options and, correspondingly, a huge range in packaging costs. How you decide to package your product is based largely on where you will be selling. If you will be selling mainly online, packaging might not be as important as when you are selling at a brick-and-mortar retail location, where you need your packaging to catch the eye of a passerby and convey the value proposition of your product immediately.

In terms of payment, for a new business working with a manufacturer for the first time, there will be some type of up-front cost, with the balance paid when the goods are finished and exchange hands. Typical splits we have seen for new businesses are 70 percent down, 30 percent when shipped, or 50 percent down, 50 percent when shipped. After you build a relationship with your manufacturer and have proven that you will pay your bills and will send a steady stream of orders, you will be able to bargain. Mina started with 50–50 terms, then negotiated for 30–70 with the 70 percent paid net 30 days (net 30 means you'll pay within thirty days), and then to 0–100, with net 60 terms. We know that among US businesses using overseas factories, there is some grumbling and negative talk about manufacturers wanting to cheat you, but in touring factories numerous times overseas, Mina also witnessed a lot of manufacturers who felt cheated by US producers who didn't pay or changed their minds and never picked up their goods. Manufacturers need to protect themselves as much as we InventHers do, but once a relationship and mutual trust are built, Mina has found that they are eager to work with you and help you succeed long term.

Questions to ask:

- What is the unit price?

- Will you provide a BOM?

- What are the payment terms?

Tooling: This is a general term used for any equipment that the manufacturer has to custom make to mass-produce your invention. No matter how simple your product is, if it doesn't yet exist in the world, it is likely that some kind of tooling is required, be it a mold for making plastics, or die-cast metal products, or a simple jig to hold your product in place while it is being assembled.

You will see that the quotes you get from different factories for tooling can vary greatly. You may get a tooling cost that is as much as ten times what you were given by another factory. Before you dismiss a factory as wanting to rip you off, remember that a lot of different elements go into estimating the cost of tooling. For example, how long does the tool have to last? This will determine the type and quality of the materials used for tooling. What is the capacity of the tool? (For example, a tool can have several cavities so that multiple units of a part can be made.)

Tooling costs are generally paid 50 percent up front and 50 percent when completed to your satisfaction (when the manufacturer has produced samples that meet your approval). To make your accountant proud, amortize your tooling costs into the unit price of your product so you can get an accurate sense of the cost of manufacturing. For example, if your tool costs $10,000 and the life span of the tool is 100,000 units, you should add $0.10 to your unit cost to make sure that your financials pencil out.

Questions to ask:

- What is the tooling cost?

- How long will tooling take?

- What materials are being used?

- What is the capacity of the tool?

- What is the life span of the tool?

Production capacity: If you think you will want to scale your business, make sure that your manufacturer can support that growth. As you will find out, shopping for a manufacturer is a *lot* of work, and the last thing you want to do is look for a new manufacturer just when you are blowing up in sales. This is why anticipating how fast you will grow (keeping in mind that this will not match up exactly with reality) is important. A manufacturer may be perfect for your initial orders of 500 units every month or so, but may not be such a good fit as you grow larger, if it simply does not have the manpower, machinery, and capacity to produce larger quantities. On the other hand, you might be drawn to a large manufacturer who also produces for well-known brands (Apple! Starbucks! Nokia!), and that fact might say a lot about their attention to quality and ability to work with US companies, but it might also mean that your order will be low on its list of priorities. We have heard of several instances where a manufacturer indefinitely delayed our friends' orders because it needed to shift machines and human resources to manufacture an order from a much larger company. So while your initial order quantity might be small, find out how the manufacturer can accommodate your needs as you grow.

Also ask whether the manufacturer will be outsourcing any part of the production to another factory (yes, your outsourced supplier may have its own outsourced supplier—that's specialization for you). For example, for the making of your revolutionary bag with the super-duper zip-on modular system, your manufacturer will most likely not want to make the zippers itself but will buy them from a supplier while it focuses on the textile and sewing portions of the product. Make sure your factory has worked with this supplier before and that it can ensure the capacity and quality of the supplier's goods are at least on par with its own. Mina had an instance where an entire production run was delayed because a new supplier to the manufacturer could not produce sufficient quantities to her standards.

Questions to ask:

- How many of machine X (whichever machine is used for your product) do you have?

- How many operators do you have?

- How many units of my product can your facility produce per day/week?

- Who are some of your other customers?

- What are the parts you will be making in-house? What will you be buying from a supplier?

Quality control: We don't know about you, but we just hate it when we buy stuff that breaks after a few uses. In fact, the poor quality of the products she was using was one of the drivers that led Mina to invent Heroclip. Most manufacturers have their own internal quality-control process, but you might want to seriously consider hiring an unbiased third-party inspection service. Regardless, it is extremely important that you give *very* detailed descriptions of what is and is not acceptable quality. It does not suffice to say, "Make sure everything lines up," or "A joint shouldn't be too tight or too loose," or "We need premium quality." These are all things that Mina was inclined to say as quality-control criteria when she started out, because, hey, can't everyone tell when something doesn't line up or is too loose? In the end, Mina's quality criteria evolved to a fifteen-page guide that specified to the millimeter what was and wasn't acceptable.

If you decide to use a third-party inspection service, they will talk to you about AQL, sampling sizes, and critical, major, and minor defects. AQL, or acceptable quality limit, is basically the worst quality level that is still acceptable. As you already know, perfection is impossible (although that doesn't stop us from pursuing it), and even the most premium product can occasionally have a defect. The AQL you set specifies what percentage of your product can have a particular type of defect before you decide that you are going to reject the whole lot. There are three types of defects in the world of quality control. The first is "minor defect." As you guessed, these are defects that are mainly cosmetic, like a small scratch, and do not affect the functioning of the product at all. A minor defect could be a smudge on the packaging or a scratch smaller than 2 mm in length. The second is "major defect," which could be a larger scratch or an assembly that is too tight (or too loose), and is a level of defect that might lead to a return from a customer but does not affect the functioning of the product. The third is "critical defect," which is something that does not allow the product to work the way it is supposed to. What you set as the acceptable percentage of each level of defect is dependent on what your product is. For a high-quality consumer-goods product, you might see numbers like 5 percent, 2.5 percent and 0 percent for minor, major, and critical defects respectively (see page 243 for testing and inspection resources).

When an inspection is done, is every single unit inspected? In some cases, yes, but for most consumer products, when thousands, hundreds of thousands, or even millions of units are being produced, checking the quality of each and every one of them is impractical. So inspection services use a globally accepted sampling method to randomly select a set number of units determined by the total number of units produced and extrapolates for the entire lot based on the number of defects, if any, found in the smaller sample.

Depending on what you are making and selling, quality is more or less important. For Mina, quality was a high priority, such that when she was shopping for new manufacturers for Heroclip, one of the things she did was take a manufacturing engineer and her product developer to tour the facility and pay specific attention to the factory's internal quality-control process. She also uses an independent inspection service when production is complete.

Shipping/freight/tariffs: An often-forgotten element in estimating costs is the transportation of your product to wherever you will be warehousing your inventory as it flies off the shelves. If you manufacture domestically, then your shipping needs will be considerably simpler than if you manufacture overseas. Large carriers like FedEx and UPS do freight shipping as well, where they will move pallets of goods rather than individual boxes (a pallet is basically a 48-by-48-inch platform on which your boxes are piled up so they can travel in a container on a boat or aircraft). It is also pretty easy to find independent long-haul trucking companies.

If you manufacture overseas, shipping will be significantly more complicated. Besides the cost, you will have to factor in the time it takes for your goods to be transported and then pass through customs. If you ship via boat it will be cheaper, but it will take at least three weeks to transport on the water and another week or so to clear customs. And then there is the time required on the front end for your goods to get from the factory to its port, and on the back end to travel from the US port to your warehouse. Even by plane (which can be up to three or four times more expensive than boat), factoring in customs and ground transport, you are looking at a week at minimum. To get on a boat or an aircraft, you will most likely be working with a freight forwarder, who basically gets bids from various shipping companies to transport your goods. For this you will need to know, at minimum, carton sizes, weight, and quantity, along with the destination and origin addresses, to get the most accurate quotes.

To make life even more complex (and expensive), if you manufacture overseas and need to import your goods to the United States, you'll also have to worry about tariffs. To determine your tariff, you will have to, with the help of a customs agent, determine into which HTS code your product falls. HTS (harmonized tariff schedule) is a product categorization system developed by the World Customs Organization that acts as common language used by all internationally trading countries. One tough thing about tariffs is that they are determined by policy makers and can literally change overnight. Mina's company had to deal with an increase of 10 percent in tariffs in 2018, which no amount of strategizing and penny-pinching could get around. If you think it might make sense to produce overseas, know your HTS when you calculate the financial feasibility of your invention. Depending on your category, it can be as low as 0 percent or as high as over 35 percent (or more) of the cost of your product.

Warehousing: Unless you have an empty six-car garage, once your business begins to take off, you will likely run out of room in your own house (we've all started by stacking boxes up in every nook and cranny of our homes). We've heard of a few instances of business owners using a self-storage unit to house their inventory, but we recommend that you set up a fulfillment system that can grow with you. The added benefit to this is that you won't have to constantly clear off tables and root around for the right-sized bubble mailer every time you have an order. You can set up your own warehousing, or you can use a third-party logistics (3PL) provider who will not only warehouse your goods but ship them out as well. There are pros and cons to using a 3PL provider. Although it is great to outsource the fulfillment of your orders, besides the warehousing cost (usually an amount per cubic meter), receiving cost, postage cost, and "pick-and-pack" cost (which will include a per-order handling fee as well as a "pick" fee for every item that goes into the order), you will have very little control over what the shipped order looks like. (Ever hear of the "unpackaging experience"?) You'll also lose the ability to put in marketing inserts or freebies on the spur of the moment—and trust us, "spur of the moment" happens a lot in a new business.

Cost of goods sold (COGS): Adding up the manufacturing costs we've described and dividing by number of units is your cost of goods sold (COGS). Yes, it is a lot. But it is also a number that is extremely important, as this is what goes out of your pocket even before you've spent a single cent on marketing or advertising. The general rule of thumb is that as your production quantity increases, your COGS will decrease because you will gain economy of scale. But as an aside, be aware that as your sales grow, costs outside of your COGS will most assuredly increase (for example, the more sales you have, the more customer service you will need to provide).

How the Heck Do I Find a Manufacturer?

As a new InventHer, you probably don't hang around with a lot of manufacturers. Don't be intimidated—you just have to start a search process. You will learn as you go!

If you are looking for a domestic manufacturer, a directory like ThomasNet.com is extremely handy. Thomas is a directory of US-based manufacturers with a nifty functionality where you can search by geography or type of product. For overseas manufacturers, Alibaba.com is a large online marketplace for suppliers based in China.

You may also be able to find some local manufacturers using your trusty research assistant Google. Conduct your search using the material your product will be made of or your product category. What is the majority of your product made from? Simple keywords like *plastic manufacturer, garment manufacturer,* or *electronics manufacturer* are always a good start. Some manufacturers, such as makers of apparel or garments, may be easier to find than those for plastics or electronics. You will have to try many search terms. For apparel, look for *cut and sew.* For wood parts or products, consider looking to the Wood Component Manufacturers Association. For metal, are you thinking aluminum or steel? Your browser is your friend.

Once you have found a few leads that seem promising, go through the questions we've outlined in What to Learn About Your Manufacturer (page 129). Also see the Manufacturer Comparison Worksheet on page 248.

If you are looking for an overseas manufacturer, there are a few options. The first is to use a trading company. These companies are pretty much turnkey: you tell them what you want, they deliver, and you pay. Sound heavenly? Yes and no. The big downside is that you are dealing with a middleman that seems to want to keep things as opaque as possible. You may have *zero* access to the factory to discuss quality control and have no clue how your product is being manufactured. All you will know is that the trading company will deliver what you want (hopefully) and charge you a prenegotiated price. This is probably a good option for products that don't really require a high level of precision and quality (many trading companies seem to specialize in tchotchkes), but otherwise, this may not be a good option. Generally, trading companies upcharge by 5 to 10 percent of what the factory charges (totally justified if they indeed deliver the goods promised), but because of the lack of transparency, it is difficult to deduce the true cost of manufacturing your product. This is an issue if you want to evolve your product or change out a part, as you are basically working blind, with little sense of what different elements of your product cost.

The second is a sourcing company or agent. These are folks who find you manufacturers overseas. Traditionally in this model, you work directly with the factories and the agents get a percentage of the value of your order. But you can imagine how there may be a conflict of interest: Is the agent really finding you the best manufacturer for your product or the one that will give them the biggest kickback? Recently there has been a movement toward more transparency in the sourcing world, with agents charging a flat rate or a prenegotiated percentage that you, rather than the manufacturer, pay.

The third model is the extended team retainer (this is what Mina uses). For a set fee each month, you get to have staff overseas where you manufacture. You get their input in sourcing as well as in managing production, logistics, and quality. If you decide that you want local "eyes and ears," it is extremely important that you get references and meet candidates in person several times, just as you would if you were hiring full-time employees. Mina's overseas team is as much a part of her company as her Seattle-based full-time employees, and she sees them in person at least once a year to make sure that everyone's goals are still aligned.

Your Manufacturing Relationship

The decision to manufacture in the United States or elsewhere is a complex one with factors beyond cost, and can depend on materials, quantity, the production time needed, and other non-production–related factors like whether your customers demand a US-made product. There are definitely challenges to choosing an overseas manufacturer—language barriers, time difference, the sheer distance that adds weeks to transportation, and intangibles like your manufacturers making products for a customer base of which they really have little knowledge. There are also many negative experiences that US companies have had with overseas manufacturers, based largely on cultural differences, which have been well documented in books like *Poorly Made in China*. In our view, while there is definitely reason to be wary of manufacturing overseas, there are also compelling reasons to embrace it. For example, cost aside, in many cases overseas manufacturers are much more skilled in what they do (Apple has a whole city in China producing iPhones).

Whether you choose domestic or overseas, you are still looking for the same things in a manufacturer. This is a relationship, and one that will last (hopefully) for many years. Put the same care into choosing one as you would with any long-term relationship. Ask yourself the fundamental questions: Do you like them? Can you easily work with them? Are they responsive? Do they ask insightful questions? Are they just giving you a price, or are they making suggestions as to how you could change things slightly to make items easier and cheaper to produce? Start with a conversation and tell them about your product. Ask them up front to sign a nondisclosure agreement (NDA) so they can't rip off your idea. If they balk at this, it is a good sign they may be difficult to work with and are not the people for you. If you show them the plans and they say, "Hey, did you think about molding this product this way?" or "If you make this piece out of another material, you can cut costs," then these are the out-of-the-box thinkers you want on your side.

If a manufacturer really wants to work with you, they may be able to brainstorm ways in which to save you money. If your product is made from plastic, for example, you have to have a prototype to get pricing on making a mold. InventHer Nancie Weston's new product (see page 145) was composed of nine different plastic pieces. One potential manufacturer showed how he could break it into three different molds (called a family mold), which actually cut her manufacturing costs in half. She knew she'd found her manufacturer.

No matter who you pick as your manufacturer, building a mutually beneficial relationship and building trust by making consistent orders and on-time payments (and sharing a drink or several) will go a long way in ensuring a collaborative relationship where the manufacturer begins to proactively initiate quality control and product development. But you have to do your part. Here's what we've learned from experience: "Consistent orders, consistent service. Inconsistent orders, no service." It's not always easy, but it is a motto we do our best to live by.

Pros and Cons of Domestic and Overseas Manufacturing

DOMESTIC PROS:

- Easier communication—no language barrier or time zone challenges

- Faster/cheaper to transport goods

- Ability to easily conduct in-person product inspections

- "Made in USA" marketing bump

DOMESTIC CONS:

- Price

- Limited availability of manufacturers, depending on your product

OVERSEAS PROS

- Price

- Lots of manufacturers to choose from

- Expertise

OVERSEAS CONS

- Capricious customs procedures

- Language barrier

- Cultural and time differences

- Need to travel far

- Slowness in shipping and sending samples back and forth, as well as inability to just drive to your manufacturer to show them what you mean

- Inability to use shortcuts when describing quality (like saying "Neiman Marcus quality," "Target quality," "Walmart quality")

Tips on Manufacturing

- Protect your intellectual property. Have potential manufacturers sign an NDA before reviewing plans. Many
  inventors opt not to send final drawings for estimates as a
  precaution. They just supply preliminary ones to provide a
  general idea of the product or send drawings of different
  parts of the product to different manufacturers.

- Know your product and the material. If it's plastic, educate
  yourself on it. Be prepared for a high-level manufacturing
  discussion. Don't allow yourself to get snowed by technical talk.

- Bigger isn't always better when it comes to manufacturers.
  Find one that will problem solve with you (a clue will be
  how engaged they are putting together a quote for you).

Nancie Weston

Grayl

WATER, WATER, EVERYWHERE

Nancie Weston had a vision: purify water anywhere in the world—a world without disposable plastic bottles. An avid outdoorswoman, she was already frustrated with cumbersome filtration systems used by backpackers and hikers. What if, she thought, I could invent a simple self-contained, ultralight water purifier in a bottle that anyone could use? She had already invented one product for the outdoor world—Sparkie, a one-handed fire starter. She crafted a few prototypes, quickly got the attention of outdoor juggernaut REI, and Grayl was born. It is a water-filtration system that functions like a French press coffee maker—potentially contaminated water is scooped up in the outer vessel; then the inner portion with a filter on the bottom is pushed downward, forcing the water through the filter and into the main drinking chamber. Clean water in seconds!

For manufacturing, Nancie went to China. There, she experienced the pros and cons of overseas manufacturing. The biggest lesson was around quality control. When she finally settled on a design and materials, she worked with a reputable factory that did work for some leading companies, like Starbucks. However, manufacturing a product like Grayl was different than churning out cups or other containers. The tolerances for deviation from the specifications were very tight. The inner chamber had to fit very snugly into the outer chamber. They discovered, through trial and error, that if the outer portion came off the mold too fast or too slow, or if it was raining and there was too much humidity in the air, the container would warp. They had to throw out thousands of units that didn't meet the exacting standards. Nancie realized that the tolerances for her product might be too tight for this facility and that a producer of medical products might be better suited for her. Nancie's contact at the new factory was excellent and had no language barrier (often a problem), but she did find that her instructions to him were often not passed all the way down to the line workers at the factory. Her fledgling company nearly went under twice as they dealt with wasted product. Ultimately, Grayl was able to get their manufacturer to correct the issues, and they were able to get a salable product.

"You've got to have quality control," says Weston. "And you have to know your product very well, and know exactly what you need. Insist they get it right."

Today, Nancie has founded a new company, Raiin, which will continue her vision for a world without single-use plastic bottles. Her new product is a household water pitcher that will not only filter out odor, flavor, and a few heavy metals like the current water filters on the market, but it will actually go one step further and remove contaminants such as chemicals, pharmaceuticals, microplastics, bacteria, and viruses in seconds. (And it's stylish enough to set on the dinner table.) For this product, she decided to have it made in the United States. She started by Googling *plastic manufacturers in the Pacific Northwest* and got a list of names. Then she started making calls. She was looking for a manufacturer who was easy to work with and large enough to grow with the company, had other well-known customers and strict quality control, and could problem solve in case something were to go wrong. She gave them a problem to which she knew the answer to see if they could think outside the box.

"I didn't want to hear, 'It can't be done.' I wanted to hear, 'Let's look at all the options and see what we can come up with,'" she said. Using the knowledge she gained with the manufacturing of Grayl, she was able to identify one that put Raiin on the path to success.

Chapter 7: Financing Options
Finding the Dollars

It's true that money can't buy happiness, but it certainly is a necessity when you want to get a product off the ground. Depending on what type of product you are trying to launch, the amount of money you need will vary, but we can almost guarantee that whatever amount you *think* you need, you will end up needing more. Again, we are not being Debbie Downers here, just being realistic. You've probably already done some Google searches to figure out how you are going to finance your endeavor, and no doubt you have come across plenty of information on sources of financing available for small businesses. There are, however, some additional considerations when you are exploring how to fund your product and your business.

Why Would You Need Money?

You might be asking yourself, Why do I need money beyond what it will take to develop the product and build up some inventory? When people buy my product, my bank balance will grow, and I can just use that money to build an even bigger inventory, sell that, make more product, and so on . . . right? Well, the answer really depends on how fast you want to grow and what your goals are for your business. Throughout the InventHer profiles you'll encounter InventHers who made different choices in terms of how they wanted to grow, which in turn affected what kind of financing they used (or, it might be more accurate to say that they decided what kind of financing they wanted, which affected how they had to grow). Michele from Excy (page 60) decided after some back-and-forth that she didn't want to take on money from outside of her team. This is called growing organically and is a wonderful way to start and run your business, depending on your goals and how you want to spend your productive time. Mina, on the other hand, decided that she wanted to grow fast and realized that she couldn't do it organically and raised money through equity investors (more on this later).

For now, let's look at all the things you might want to do that cost money. Whether or not you think you need to worry about money, it is vital to think through the various elements listed in this chapter as you decide if you want to make a go of your idea. Note that to truly build a financial model, you will also need to estimate how much you think you will sell. Of course, how much you sell will depend on what resources you have available to market and sell your product, and the resources you have will be affected by how much you sell—it's the age-old chicken-or-the-egg story. Here is just a quick peek at some of the things that, beyond your prototype, will cost money.

PRODUCT AND INVENTORY

The first thing, of course, is the product itself. As we discussed in Chapter 6: Manufacturing (page 127), your cost of goods is more than what your manufacturer charges you per unit. It is also the tooling involved, the freight, the quality control, the warehousing, and the tariffs (if you manufacture overseas). It is easy to forget when you are starting out with just a sewing machine, doing all the production yourself with some discounted fabrics, that as you grow, you will have to buy more sewing machines, buy more fabric, get additional help to package your products, pay for storing all your stuff, and so on.

The second instance that necessitates more money is if your invention is more successful than you had even dared to hope. Let's say your wildest dream comes true and Costco orders 500,000 units of your self-cleaning lunch box. They want them in six months, and they will pay you sixty days after they get the shipment. Each lunch box costs $10 to make (sounds pretty high, but let's keep the math easy), so making 500,000 will cost $5 million. Your factory wants half the money to start the order, and half the money when they are done with the order. That means you have to come up with $2.5 million just to get going and another $2.5 million soon afterward, *months before* Costco will pay you!

This scenario where you have to shell out money *waaaaay* before you get paid is called the working capital cycle, and it is one of the biggest financial challenges that a physical-goods businesswoman faces. You can have the bestselling product, but still be staying up all night wondering where the next dollar is going to come from. In fact, as illustrated with the Costco / lunch box example, the faster your business is growing, the more money you need, and this is because of the gap between when you *pay* for your goods and when you get *paid* for your goods. If you are growing quickly, then the problem is even bigger, because every time you order inventory, you have to shell out a larger sum of money in anticipation of higher sales. It is extremely important that you grow at a pace you can afford, or conversely have enough money to grow at a pace you want. "Growing out of business" is really a thing; many successful entrepreneurs live in constant fear of this, and we don't want it to happen to you.

MARKETING

One of the other key things that can cost money is marketing. As we discussed at length in our chapters on distribution and marketing, the fact is that customers have so many options in terms of what they can buy and where they can buy that if you just sit around waiting for people to come to you, you will be lonely for quite a while. You need to get out there and *tell* people about your product. This can cost money in the form of designers, videographers, copywriters, digital marketers, and Amazon specialists.

EMPLOYEES

When Mina was working out of her mother-in-law apartment among piles of boxes, it was difficult to imagine making enough money to hire employees. But as sales started coming in and she began receiving positive feedback from customers around the world, she realized that she was maxed out. It is hard to be at a point where you have too much business for you alone to manage but not enough to justify a full-time person. Whether you start out with part-time or full-time help, and whether the help comes in the form of your grandmother or a corporate dude in a suit, as you grow you *will* have employees, and they will be essential not just to manage the business you have at the time, but to grow the company much bigger than it is. Although new businesses start with one or two employees who basically do everything that the business needs, as you grow you will want employees who are highly experienced and can lead a team in a specific area. These folks, who are at the top of their game in their field, don't come cheap. As a result you may find yourself in a position where your payroll to staff the people who will help you take your company to the moon and beyond is greater than what your current sales bring in. This is when you need some form of financing.

PROFESSIONAL SERVICES

Although there are many DIY software programs out there to help you do all the not-so-exciting-but-necessary things to start and run your business (legal! accounting! bookkeeping!), there will come a time when you need more personalized expertise than what a software algorithm can give you. An obvious example of this is when you are patenting your product—you want a great attorney who can write a compelling patent application so that it is approved (and before that, advise you on whether your idea is likely to be approved). But even something as straightforward as bookkeeping can become complicated quickly, with various payment terms, multiple purchase orders to your manufacturer, and so on, such that you might want to hire a professional rather than continuing to do these tasks yourself.

Some specialized services may come in the form of onetime help rather than on an ongoing basis (like bookkeeping). An example of this is a product developer or an engineer who might come on board for a short period of time to give you advice (or actually do the work) on a very specific problem you are trying to solve.

It is critical to get reliable, ethical professional-service providers in your corner once you decide to make a go of your invention. Until then, be scrappy and get free advice from your friends, or friends of friends, or friends of friends of friends, but remember to take any free advice with a grain of salt: this is an area where you can get what you pay for.

Your Financial Projections

To figure out how much you will need to get to a specific milestone (selling your first product, for example), you will need to do some projections of estimated income and expenses. Check out our sample profit and loss statement (at right) to help you think about the categories of expenses you will need to estimate. With any spreadsheet program, you can easily create and examine different hypothetical scenarios of varying revenues and expenses. Create your own profit and loss statement by photocopying the worksheet on page 251.

A Sample Profit and Loss Statement

Disclaimer: We are not accountants! These numbers are totally made up.

Income			
	Revenue		$1,000,000
		Discounts	$50,000
	Net Revenue		$950,000
Cost of Goods Sold (COGS)			
	COGS		$400,000
	Total COGS		$400,000
	Gross Margin		58%
Operating Expenses			
	Research & Development (R&D)		
		Product Design & Development	$50,000
		Prototyping & Testing	$7,000
		Tooling	$20,000
		IP Protection	$10,000
		Total R&D Expense	$87,000
	Sales & Marketing		
		Content Creation	$5,000
		Advertising	$50,000
		PR	$24,000
		Social Media	$12,000
		Trade Shows	$20,000
		Customer Service	$12,000
		Total Sales & Marketing	$123,000
	General & Administration (G&A)		
		Payroll & Benefits	$100,000
		Software & Equipment	$16,000
		Professional Services (Legal, HR)	$50,000
		Insurance	$5,000
		Meals & Entertainment	$10,000
		Office Supplies	$2,000
		Bank Fees	$2,000
		Occupancy (Office Rent, Utilities)	$30,000
		Travel	$5,000
		Total G&A	$220,000
	Total Operating Expenses		$430,000
Earnings Before Interest, Taxes, Depreciation, and Amortization (EBITDA)			$120,000
	EBITDA Margin		13%

Sources of Money

The InventHers we know are nothing if not tenacious and determined when it comes to finding ways to bring their inventions to market. From maxing out credit cards to cashing out retirement funds (what's a better investment than yourself?) to keeping a day job to pitching a hundred investors in a year, there are as many ways to fund the commercialization of your invention as there are, well, inventions. In our InventHer profiles, you will see just how wide-ranging this is, in terms of how much money is used to start a business (start-up capital) and where the money comes from. (Hint: It ranges from thousands of dollars to $8.5 million, obtained from savings to venture capital firms.) Since we don't know your particular financial situation, we will cover the main sources of funding that you will come across.

BANK LOANS

Traditional bank loans are not always right for new businesses because they tend to require twelve consecutive quarters of profitability. If you miss one, you have to start all over again. It is possible that you are one of the lucky InventHers who doesn't need financing until you meet this criteria, but for most, bank loans may be a classic example of "best when you don't need it." No matter how nice the small business rep at your bank is and how sincerely she says she loves local, small businesses, all banks are guided by federal regulations and you will probably not be able to get a significant line of credit or a loan initially. Depending on your credit, however, you may be able to get a small line of credit (with a personal guarantee), which is helpful but may not be a game changer for your business.

There are shelves of books written on this topic alone, but here is the gist of it. There are individuals called angel investors: equity investors who will give you money in exchange for some percentage of your company. Of course, they will have expectations of your growth and your ability to generate a financial return for them, whether through a payout when a larger company buys your company or through profit sharing or dividends once you have enough profits to share. There is no such thing as a free lunch (even from angels).

Unlike venture capitalists, who invest other people's money in companies as a service, angel investors are generally wealthy individuals who invest their own money in companies. The term *angel investor* apparently originated in the theater community when wealthy theatergoers would give money to flailing Broadway shows to keep them from closing. Angel investors must have at least $1 million in assets to be considered "accredited," so there exists a huge range of potential angels in terms of how much they invest and how often and basically how much discretionary money they have. If Bill Gates invested individually, he would be an angel investor, the same as your next-door neighbor who, despite never having made more than $35,000 per year in his entire work life, has assets of $1 million thanks to frugal living and good fortune in the stock market. These two angel investors would most likely invest different amounts of money in vastly different numbers and kinds of businesses.

In general, angel investors come in early in a company's life. You may have heard of some software firms that get millions of dollars thrown at them with just a prototype (or even with some scribbles on the back of an envelope as the dot-com-boom legends / cautionary tales go). That most likely will not happen to you (nothing personal, but raising equity funds as a physical-product company is harder). There are, however, many angel investors who are enthralled by physical products, especially if the products are patented (or patentable) and already have a track record of consumer demand. Most companies, when they are raising funds from angels, are not looking for huge amounts of money, at least by investment standards. Mina started with a round a bit over $500,000, raised from about fifteen people. In her second round the next year, she raised over $1 million from another fifteen. She was able to raise more the second time around because of the track record she had built over that year.

The chunk of your company that your angel investor gets depends on two things: (1) the size of their investment, and (2) the valuation of your company. Note that valuing a young company is *really* hard. It is not just about the sales and performance you currently have, but also about your *future* performance. And since nobody can see what will happen in the future, for the most part your valuation will come from comps. Just like when you are buying a house, comps are comparable companies that already have valuations. It could be something as casual as, "Oh, this company out of Bend, Oregon, had a physical product serving a similar market with similar revenues, and its valuation was $3.5 million, so your valuation is somewhere around there." Or it could be "A company in this industry is generally valued at X times its previous year's revenue." And people will disagree wildly on your valuation. When Mina was raising money, what her investors thought her company was worth varied by as much as $3 million. In the end, you will be the one setting the price on your company with the advice of people who know your industry and the start-up scene really well. Those folks you sought out earlier for professional help will come in handy now. The amount of equity an investor gets in your company will be *proportional* to the value of your company after you have finished raising your money. The higher your valuation, the less you give away, of course, but remember that if you are the only one who thinks your company is worth $10 million from the get-go, you may have a really hard time raising funds.

Let's talk about that other kind of investor: the venture capitalist. They are not often applicable for new inventors, but you should know about them. Until your company is large enough that your valuation can support a raise of $5 million or more, there may be little use in pursuing them, although it is always nice to start building a relationship for the future or for perspective. We're big fans of getting perspective from people in the know. Always be networking and learning, but don't spend a lot of time seeking out venture capitalists at this stage of your game.

Here's the thing. Raising money is not a completely rational exercise (perhaps the understatement of the decade). Just as people buy all kinds of stuff for reasons beyond the function and value to their lives, people will decide to buy a chunk of your company for all kinds of reasons, some based on rational analyses of the opportunity and some that have nothing to do with you or your business. Some of the more rational reasons for investing in your company may be things like your past performance (not just with this product but your previous jobs or businesses), how compelling they find your product (i.e., would they or people they know use your product?), how other comparable companies have done in terms of growth, and what they think of your industry. On the flip side, they might turn you down for equally rational reasons, even if you disagree (e.g., "I think you are amazing and are doing wonderfully, and the product is awesome, but I just don't like where I think your industry is going" is something that you might hear quite a bit). But when it comes to super-early-stage equity financing, there are many more reasons that have nothing to do with you as an InventHer or your product or even the industry you are in. Investors may not decide to invest in your product or business because their friend did or didn't invest, they invested too much already this year, you remind them of a girl that didn't give them the time of day when they were in high school, or they were distracted when you were giving your presentation and totally missed all the reasons they should invest. The reasons are many and varied and sometimes infuriating. For example, even in this day and age, being a woman can also be a reason for not getting investments from certain, shall we say, "backwards" individuals, as several of our InventHers featured in this book found. Sadly, the patriarchy is real. The struggle is real. But so is the sisterhood. And tenacious InventHers eventually find their tribe.

ALTERNATIVE LOANS (SHORT-TERM, REVENUE-BASED)

Given that equity financing is not always desirable or feasible and that traditional bank loans are almost never available to new inventors and businesspeople, enterprising individuals have come up with alternative ways to make money available . . . at a price, of course. These entities are willing to lend you money quickly with only minimal financial background checks, but at exorbitant rates, like 1 percent *per week*. If you annualized it to an APR (annual percentage rate), it would be 52 percent! When Mina raged about this ridiculous rate to the poor guy who cold-called her with the offer, he pointed out that these loans are intended to be very short term (a month to a few months) to cover urgent expenses (or to pay for opportunities) and are not meant to be long-term situations where the business owner spends all her revenue on interest while the lender makes a killing. We don't recommend it, but it's an option.

Another new type of loan that has become more available in recent years is a revenue-based loan. Basically, you get a loan that you pay back monthly based on your revenue that month. They call this an entrepreneur-friendly loan because when sales are down, you also pay down less, and when they are up, you pay more off. In researching this, we found that many of these lenders want a guaranteed return on investment—meaning you cannot pay off your loan too fast. Also, most seem to work with software companies that make money on subscriptions, which are more predictable. However, with the prompt cash cycle of e-commerce, these revenue-based companies are seriously looking into making loans available to physical-product companies based on monthly e-commerce revenues. Check in with your friend Google to see what options are available when you are ready for financing.

 HOT TIP: If you get a loan, make sure you can pay it off early without penalties!

CROWDFUNDING

We have a soft spot for crowdfunding. As you already know, Mina started Heroclip as a crowdfunded company, and Hilary has worked on numerous crowdfunding campaigns as an outside consultant. We like to think of crowdfunding as a presales and marketing tool rather than a capital-raising tool, although you certainly will raise capital. Started as a way to pool resources from supporters for special projects, crowdfunding has become a major launch platform for new products, and there are now well over a thousand crowdfunding platforms throughout the world (check out the Resources section on page 233 for some big ones).

The beauty of crowdfunding is that you get to sell your products *before* you manufacture them, and the buyers are willing to wait months to get their purchases. In exchange, you are expected to give a substantial discount, which, in our opinion, is well worthwhile. This is a godsend for a new InventHer, because not only can she get a good sense of what kind of demand her product will generate, she will also have the money to get these products manufactured. This is assuming all goes well, of course. If you don't reach your project's funding goal, supporters are not charged and you're not obligated to produce the product. Even in the worst-case scenario—not reaching your goal—you will at least have one data point that gives you an indication for how popular your product will be. But crowdfunding requires a lot of resources, especially since so many products are now launching on these platforms. Just like you will want your product to stand out from all the other goods competing for attention at Target, you will need to use all your marketing knowledge and skills to rise above the rest on crowdfunding platforms. Mina has done this three times, and each time she sold a higher quantity of whatever she was launching, but she also spent exponentially more time and resources. Still, we think it is still an incredible way to launch your invention.

A Final Word

You *will* need more money than you thought you would. Getting it will be hard. But keep your eye on the prize, because nothing compares to the feeling you will get when your product is all over the world and customers are writing and telling you how it saved the day for them.

Amy Buckalter
Pulse

FINANCIAL PREPARATION, LIKE FOREPLAY, YIELDS RESULTS

When Amy Buckalter entered menopause, inspiration came with a cold jolt. Personal lubricant: Why is it so cold and messy? During a doctor's appointment or an intimate moment, who wants an icy shock of unpleasantness? Her research revealed that lubricant was ripe for reinvention—formulas were old and filled with unnatural and toxic ingredients. The idea for Pulse was born: an elegant, warming, and hygienic delivery system for a better personal lubricant experience. Think of the Keurig machine and pods, but for lube, not coffee.

Amy was not an inexperienced player. She'd already built a career developing outdoor and sport-related brands. She'd taken major brands like K2 to the next level, then leveraged her experience into a lucrative consulting role. Now she would bring all this know-how to Pulse. She did enormous preparation to lay the groundwork for success. The first task was market research. Amy hired a researcher with a PhD in cognitive psychology and human factors to conduct interviews with nearly four hundred women and validate the need. Her market turned out to be postmenopausal women, millennials, and gay men. When the research indicated that she had found a significant unmet need, she began to build her team, with the right expertise, to take Pulse to orgasmic (sorry) levels of success.

Amy set up a matrix with the functions she needed to take a complex product like Pulse to market. Product-development experts, sexual-wellness experts, legal experts, marketers, people with expertise in personal care, people with expertise in medical devices, and more. She didn't have money to hire them all, but was able to bring people on as strategic advisers to help her figure out what she didn't know. She discovered that she'd need FDA approval, which requires time and money. She found a key player early on with Dr. Lilac Muller—an MIT-trained product-development and engineering executive who began her career at NASA.

With her team assembled, Amy could create a detailed business and finance plan, knowing what it would cost to take Pulse to market. This included, from her experience, the "shit happens" fund to correct "the shit you can't plan for." (A wise fund for any entrepreneur to have!) She knew she'd need a significant amount of capital to get up and running, and didn't want to waste time going in front of many different angel groups for small amounts. She initially put in $100,000 of her own money and set a high minimum investment amount. It took eight months to get ready, but the preparation paid off. She raised $1.2 million in just eight weeks. With this money, she was able to start the development process of complex designing and prototyping.

Then she brought the investors back in to show them her progress and raised another $2 million in twelve weeks. FDA clearances, patents, new device designs, and new lube formulations continued. Amy's funding prowess paid off in spades—at this point she's raised more than $8.5 million between 2014 and 2018, and Pulse began generating sales, first online direct to consumers, then through Amazon. It's been featured in *Cosmopolitan*, BuzzFeed, *InStyle*, and many other media outlets, and is a darling of the emergent femtech industry. The growth of Pulse sales continues to impress: according to Amy, from 2017 to 2018, sales of the Pulse dispenser were up 160 percent and the Pulse pods (which are usable without the dispenser as well) were up 634 percent. Definitely sexy numbers.

Amy's approach to investors and fund-raising, besides being meticulously planned, was based on relationships. "I really wanted to get to know them, have them get to know me so they can have a high level of confidence in my leadership and work ethic, and have time to give them a full understanding of the potential of the venture, not just a ten-minute pitch. There are many we don't want in our investor family or on our journey. I feel personally responsible to get these people to the promised land of investment success. Sixty-five percent of our investors are women. They include my college roommate, a graduate school friend, ex-relationship partners, people throughout my life who have known me both as a person and a professional, " she says. "We're on this journey together."

Chapter 8: Building Your Team
Swipe Right for Your Business

One of the scariest and most challenging things you'll have to do as you build your product and business is get help. Even if you are the superwoman we know you are, if your product is successful, it will at some point become impossible to manage on your own. One woman can't do it all—you will never have the time, the expertise in certain areas, and even the interest to do some of the tasks your business requires. (We're looking at you, bookkeeping—we'd rather be doing *anything* else.) So you are going to need other people to fill in some gaps: you're going to need a team.

In our view, the composition of your team is probably the most important thing that determines whether you succeed. Have the right team members, and you will soar high with the knowledge that they have your back as you reach for the stars. Have the wrong people, and you will quickly exhaust yourself physically, mentally, and emotionally from dragging them along as your company struggles to get on the same page about the simplest tasks. And when we say "team," we are not just talking about the employees on your payroll who help take care of daily operations. Building a successful business around a product can be so complex we consider part of our team anyone who can materially affect how our business grows: our full-time employees, of course, but also our part-timers, contractors, bookkeepers, accountants, lawyers, investors, outsourced fulfillment centers, and even shipping carriers. We'll look at employees here.

Hiring Employees at a Start-Up Is Different

Go on LinkedIn and you will see any number of recruiters lamenting that, due to the extremely low unemployment rate we're currently experiencing, finding candidates to fill jobs is like pouring water in a bucket with a gigantic hole in it. They say that candidates will take a job only to leave when they get an even better offer, if they return recruiting phone calls in the first place.

As hard as it is for a larger established company to find employees, it is even more challenging for you as an InventHer with a start-up. But don't despair; it's not impossible. Let's look at the challenges you might encounter when building your team and what you can do about them.

 HOT TIP: Your team is not just full-time employees, but anyone who can materially affect how your business grows.

Money Can't Buy You Love, But Will It Buy You Your First Employee?

The first and most obvious challenge you may face as a start-up is that you don't have the payroll budget that Microsoft and Amazon do. You also don't have these companies' recruitment machines that can process and pore over thousands of résumés to find the two or three candidates with the perfect qualifications for the job. Finally, you can't offer the daily free lunches, on-site dry cleaning, $1,000 ergonomic chairs designed by famous architects, shiny windows framing an inspiring nature scene, free gym memberships (and trainers!), matched retirement plans, and subsidized in-vitro fertilization. Yes, some of our "corporate" friends have all of these bennies and then some. But that's OK. You've got something they don't have.

Offering lower pay and fewer benefits can often result in slim pickings in terms of potential employees. There is the person who is supremely unemployable for some reason and is unable to get any other job. There's the person who has left an established company job to "figure things out" or "start their own thing" and is looking for a job to fill the gap while their mind is on something else. There are the very young or inexperienced people with zero work history. However, there is another elusive category: the supertalented person willing to forego higher pay and fancy perks because she is on board with your passion and vision, wants a cool and interesting work environment, wants to learn and is willing to put in the sweat equity to be part of something . . . and who might also be motivated by the possibility of equity or stock options.

You can probably guess which category to focus on. While many new business owners will say, "We can't afford to hire great, committed people," we like to say, "You can't afford *not* to hire great, committed people." (Not that we haven't made our share of hiring mistakes.) A large company with several people performing the same work won't stop operating if one or a few of them are not quite so motivated, talented, or generally on top of things. They have other folks who can cover for them (intentionally or not), and the company goes on. A small company like yours doesn't have multiple people doing the same work, but the opposite: one person doing multiple jobs. If this infinitely important person doesn't have the skill sets they need to perform their job(s) or is simply lacking motivation, you have a major problem.

As you have begun to see in this book, bringing a product to market requires your left brain, your right brain, and everything in between. It requires that you are as gracious about picking and packing products to ship as you are meeting with your first investor. In our experience, the most important qualities of a first hire are a love for learning new things and the ability to wear different hats as needed. Since there is no six-month onboarding and training program at your company and you probably already have your hands full doing whatever needs doing, employees must be proactive about asking questions, detecting problems, and finding creative solutions, rather than waiting for you to direct them and hold their hands. Oh, and it helps tremendously if they love to work (because there will be an infinite amount to do) and don't have an attitude about doing things that might seem outside or below their qualifications. Once, the whole Heroclip team had to go to the warehouse to open the packaging on 10,000 units after learning that there may have been cockroaches in the shipment (there weren't). Yeah, we don't ask for much.

So how do you find these mythical creatures who love to work, don't mind doing dirty work, are whip smart, will take lower pay, and will basically commit to *your* company like it is *their* company? To answer this, bear with us as we rely on some well-known sayings and offer a couple of our own maxims.

Just Say No

In our experience, saying no to job candidates is as important as saying yes. It is sorely tempting when your energy/time/cash tanks are all empty to set the pay super low and hire the first person who responds to your job posting. "Who cares that they have none of the skills I need at my company? I'll work with them and teach them and watch them so they can't screw up." Don't do it! Just say no until the right person shows up. It is tremendously time-consuming and energy depleting to guide and monitor someone 24/7 (as anyone with a young child knows), and as an InventHer and now business owner, you already have an overflowing plate and no room to be constantly worried about whether your new employee knows what she is doing. This is not to say that your first employee has to have every skill you need. It means she should definitely have the skills to do the one or two things that you absolutely cannot do yourself, along with being smart and interested enough to learn new skills as the needs arise. It also means that she is driven by the prospect of having real impact on a company.

If you are unable to find a great candidate at the pay rate you are offering, wait. Continue being a one-woman show until you can pay more. Trust us, working on your own is twice as productive as having an employee who needs supervision and guidance all day long or is simply incompetent or unmotivated. In case we sound like we don't care about employee development, we do! We really are nice people and want our employees to grow as people and as professionals. However, the first employees at the initial stages of your business need to not need you so much, so you can focus on growing the company.

Give Them "Skin in the Game"

Remember how we said that you want your employees to commit to your company like it's theirs? So, let's make them owners (or at least future owners). While your cash situation isn't anything that will put you on the *Forbes* wealthiest list quite yet, you do have plenty of shares of your company. If you have successfully conveyed how your product is going to sweep the nation and how you are looking to build a team that will help you grow your company and change the world while you're at it, an enticing compensation tool may be equity or stock options in the company (see page 180 to learn more). The idea is to have the interests of the employee aligned with your own interests, so you can work toward a common goal, for the well-being of the company.

"Oooh," you say. "What if I give them extra skin in the game and hire someone just for a big chunk of the company and no cash? Won't they be even more motivated than if I just give them a little bit of the company?" Let's take a pause here. While this might work for lucky folks who are independently wealthy and just want to work with you because they believe in your vision and the potential of your business, the fact is that most people do require a paycheck, and those who will work for no cash and all equity will likely have a day job that provides this paycheck. And while it is theoretically possible for people to work nights and weekends and make effective team members, it has been our experience that the job that doesn't pay cash tends to take second chair to the job that does. When it is six o'clock and time for her to complete an urgent project for you, then an equally urgent project comes up at the day job, guess which one will see the light of day? Yep.

Hire Slow, Fire Fast

This is a common mantra in the start-up world, and one with which we agree completely. As we said earlier, it is tempting to hire fast when you need help yesterday—never mind that you have no time to read a hundred résumés when the first one you see seems to be just fine. However, if there's one thing we've learned the hard way, it's that hiring slowly, getting to know your job candidate as well as you can, and having her get to know you, is the only way to go. Hiring slowly to us means meeting with the candidates multiple times. We know you're busy, but you will be spending more time with this person than with your spouse, and you wouldn't marry someone after one date, would you? Take her to lunch and see what she is like in a social setting. Call every reference she lists and ask how she did on projects with very few guidelines. And while you should portray employment with you and your company as positively as possible, make sure you don't pull any punches when setting expectations. Tell her the first priority could be one thing now, but a completely different thing in an hour. Tell her that she's expected to be a 24/7 salesperson and ambassador for your product. Tell her that the water pipes at the office leak and you're both responsible for emptying the bucket. If possible, have your trusted advisers (friends count here) meet the person and get their opinion on how the two of you will do in your "marriage."

If possible, hire candidates for small projects before hiring long term. Mina had some epic hiring fails before giving candidates small tests and projects to gauge their skill level and problem-solving abilities. When Hilary was looking to hire a graphic designer for a client, she gave them multiple small projects first. While a portfolio can tell you their skill, it is different than working with them, seeing how they take feedback, and noting how quickly they grasp your ideas and vague directions. After working through a few projects, Hilary could tell who could handle the "I need this now!" and "Can you make it more cartoony but not a cartoon? Like Japanese anime but more adulty . . ." or similar feedback. (You must pay for these "tests"—it isn't ethical [or legal?] to ask candidates to "audition" for free.) The nature of the projects depends on the position you are filling, but even the most mundane exchanges you have with your candidates can give you a good sense of how they approach problems and handle things that are new to them.

Another tough challenge: the breakup. Even after all this slow hiring, it may turn out that your employee is not a good fit for the company—sometimes something you never thought to find out during the interview process turns out to be an issue. As an extreme example, at Heroclip, it was discovered that a hire who came highly recommended by a trusted adviser and who had a solid work history was taking pornographic photos of himself during work hours and saving them on his work computer. Upon further investigation, they discovered that he was also using work time to peruse and post on porn sites. An awkward and necessary conversation and firing ensued. It's drama no one needs, but such things do happen, and it will be up to you to fire fast.

As painful as it can be to let go of someone you have worked closely with, there are many reasons for acting fast once you have decided that an employee will not work out. Besides the expenditure of financial resources on someone who is not adding great value to your company, keeping someone on when it is evident that she is not a good fit is an emotional drain (again, no one needs additional drama or turmoil). More importantly, an employee that is a poor fit can have widespread impact by demoralizing your team and promoting a feeling of unfairness if the team is forced to pick up the slack from the employee who's not pulling her weight. Of every business owner we have spoken to about this, not a single one regretted firing an employee—they only wished they had done it sooner.

 HOT TIP: When screening a potential candidate, check out their social media profiles and see what they are posting.

You Can't Grow Without Changing

Besides the fact that you have less time and money when hiring as a small start-up, there's another reason you are different from the larger, more established companies of the world: While large companies like Microsoft or General Electric (GE) are lucky if they grow 10 percent in revenue year over year, your company may be growing faster than the proverbial beanstalk, doubling, even tripling and quadrupling in the first couple years of its life. As a company's revenues grow, the team size grows, and individual roles within a team change. One adviser we know says that every time a business doubles in revenue, it is like building a whole new team. Everything changes, and those who can't adapt must be left behind. What this means is that the person who was perfect as the right-hand woman (or man) when you were a two-person team may not be such a good fit in the same role on a ten-person team.

When you are a two-person team, what matters most is that each person can wear multiple hats and switch gears quickly and efficiently. As you grow, in our experience, what becomes more important is deep expertise and experience in a specific functional area (like sales, marketing, operations, finance) that the team member can "own." In other words, instead of having ten people who are willing to do anything, we need ten people who are really, really good at one or two things and have the leadership skills to run their own departments.

As the needs of the company evolve, it doesn't mean you ditch your early employees and hire new ones ASAP. Letting go of employees may need to happen, but motivated, committed team members are super rare (recall that we refer to them as mythical creatures). If she is willing to grow professionally and is able to rise to the occasion, investing in an early employee's professional development to enable her to become a leader of a functional area might make a lot of sense.

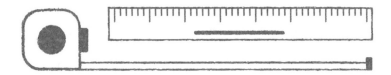

Frequently Asked Questions

Hiring can be a tough process with a lot of different things to consider. After "I have an idea for a product, what do I do now?" (the response to which is this book), the questions we get the most as mentors have to do with building a team. Here are the most frequently asked questions, and our answers.

I can't afford a full-time person, but I need help! I'm working so much my kids are beginning to refer to me as that lady who microwaves food for us. What do you think about hiring a part-time person? And what's the difference between an employee and a contractor?

We are totally in favor of starting by hiring a part-timer temporarily. In fact, we encourage it, because by the time you are ready to hire full-time, you will really know that person, the quality of their work, and their work ethic (remember our advice to hire slow). In the long term, though, we prefer full-time employees. When someone is working only limited hours, it is difficult for them to be in the loop about all the things that are going on at your company, which could lead to inefficiencies and mistakes.

As for part-time employee versus contractor, the term *contractor* doesn't necessarily mean part-time work. Working as an independent contractor means that they are responsible for paying for their own taxes while you pay them a gross wage based on hours worked (for employees, as you may know, we need to deduct myriad taxes and then pay the government on their behalf). But beyond who pays the taxes, there are a number of subtle differences, such as who provides the equipment they use for work, how much direction they receive, and so on. In the end, the employee versus contractor question is dependent on exactly what work they will be performing and under what circumstances.

What should I pay my first employee?

First, do a bit of research to see what others in your area are paying someone who is performing a similar role to what you need. You can get this data by asking other business owners and by going to sites like Salary.com. Generally, when you get a range of salaries, you can tweak your offer based on the specific qualifications you're seeking in a candidate or the circumstances of the job. If you are not able to quite reach the range of what is generally paid in your area, consider offering other perks like a flexible work schedule, the choice to work remotely, a 4/10 workweek (meaning four ten-hour days per week, rather than the standard eight hours for five days), the ability to bring their pets (or kids) to work, and so on. We have discovered that while the salary is important and we would never encourage anyone to exploit somebody, nonsalary perks can be quite enticing to today's workforce. And if you feel like this person is someone you want as a shareholder of your company, consider equity or stock options.

Do I need a cofounder?

There are some supersmart people who think that you cannot have a successful company without a cofounder. Their reasoning is that you need someone to bounce ideas off of, to provide expertise you don't have, to keep you from having tunnel vision, and that this cofounder has as much invested in the company as you do. And you *do* need all of these things, but we would not go so far as to say you *need* a cofounder to succeed. In fact, a recent TechCrunch study showed that 52 percent of the companies that successfully sold to larger companies (a goal of many entrepreneurs) were started by solo founders.[15]

So, our thought is that if you have a person you absolutely trust who can take joint responsibility for starting the company, by all means go for it, but don't hold back because you don't have such a person. Here are a couple tips if you *do* have a cofounder.

1. Make sure the division of roles, responsibilities, and decision-making authority is *very* clear. Do *not* attempt to do everything by consensus. This will *not* work long term. *Every* project should have *one* and only *one* owner who is held accountable and is responsible for seeing things through. Maybe you have the final say on finance-related things and your cofounder has the final say on marketing projects, but *one* person should be the lead. This doesn't mean you can't solicit advice and feedback from the other person, but ultimately it is the project owner's call. Too many teams with cofounders attempt to "co-do" everything, which ultimately leads to endless discussions, diffused responsibility, inefficiencies, and lack of accountability.

2. Remember that cofounders are difficult to fire, and we *strongly* advise you to seek legal counsel when you are drafting your corporate documents and anything to do with this cofounder, as things (including feelings) are known to change. We have heard far too many horror stories of cofounding situations going sideways for all kinds of reasons.

What's the difference between equity, stocks, shares, and stock options? And how much do I need to give?

Equity basically means ownership in a company, as do *shares* and *stocks*. If you have shares/stocks/equity in a company, you are an owner of the company. There are also preferred and common stocks, which have different rights, so make sure you work with an experienced attorney when drafting up stock agreements.

Stock options give you the ability to decide at a future date whether you want to buy into the company at a preset price. The idea is that your company will grow and its stock price will go up, but you get to buy it at the price it was when the options were granted. Let's say in 2019 a share in your company is a dollar. You issue options at this price. In five years, each share is worth ten dollars. Whoever received stock options at the dollar price gets to buy the ten-dollar share for a dollar. Nice, huh? Yes, it is. It is so nice that you need to put a vesting schedule in place so that someone who works for your company for just a few weeks isn't in the unfair position of being able to get shares for cheap in the future. A vesting schedule specifies that the option grantee will receive the options in installments (usually over four years). Consult your attorney!

Do I need an office? Or should everyone just work remotely so we can save on rent?

We love saving money, and having a home office is a great way to achieve that goal (ask your accountant about writing off a portion of rent or mortgage payments). And there is a lot of research that shows working remotely increases productivity. For us, when we are together in the same space, communication is easier and we come up with more creative and better ideas. This doesn't mean just at in-person meetings, but even during those unplanned moments when you and your teammate happen to be waiting for your lunches to heat up. To test out how your team works together, join a coworking space and work together a few days a week while working remotely on other days.

Is it OK to hire close friends and family?

That is a tough one. We know a lot of people who work very well with their family members, and we also know a lot of other people who don't. When you're starting out, it's easy and reassuring to rely on the folks you trust, folks who get you and believe in you, and those are your family and friends. We think the key to success in working with your close circle of friends and family is to set clear expectations about their roles, just as you would with cofounders. This becomes even more important as your company grows and you hire more people—the last thing you want is to foster resentment among your nonfamily employees who think that your family members are getting preferential treatment or are unduly influential. And even family and close friendships can go south, especially when money and company shares are involved, so make sure you consult an attorney who can outline contingency plans.

You talked about firing fast. How do I know when my employees are keepers or should be cut loose as we grow? I mean, I like them and all, but it is hard to tell if and how they are contributing.

As a new business, you are so focused on growing, selling, and creating that the last thing you might want to spend time on is reporting.

But as we have learned over the years, reporting and tracking—along with goals and KPIs (key performance indicators)—are critical not just to employee performance but also to employee happiness. We think that people inherently want to do well in their jobs, but often, especially in a start-up, they don't have a strong sense of what is truly important. Give your employees the comfort of knowing exactly what their goals are (which you will develop with them) and how their performance against these goals will be measured. We like numbers, and we like setting goals like "Increase traffic by 10 percent month-over-month." But more aspirational goals like "Learn how to make videos using green screens" are important too, because they show your employees' ability to be proactive about their own learning, especially as it relates to your company. Have your employees track and report their performance monthly, and do a review together to learn what enabled them to reach or exceed their goals (or what led to them coming up short). Figure out what you can both learn from the past month's performance to do even better.

Besides performance, we think an important deal-breaker is lack of cultural fit. Even when your team is very small, you will have values and norms and practices. Are you the kind of company in which team members can freely and informally give their opinions, or one in which there is a formal time, channel, and process to give feedback? Are you the kind of company where the big picture is the only important thing, or one where every dotted *i* and crossed *t* matters? Do you believe that good performance comes from fostering intra-team competition, or is your philosophy that working together can improve individual performance? A poor cultural fit is miserable for the employee and worse for you and the team, so when evaluating whether someone is a "keeper," make sure you consider how your employee fits into the team, in addition to their work performance.

My just-out-of-college employee thinks her title should be COO. It's no skin off my back. Should I say yes?

In the start-up world, there is often what we call *title inflation*, where people with little experience get big titles. The reasons vary. Sometimes entrepreneurs think there are certain positions that need to be filled for their company to be considered legitimate (the popular ones are CEO, COO, CMO, CTO, and so on). Sometimes the title inflation occurs to keep a new hire happy and is accompanied by thinking along the lines of *I don't care what title they have; they'll still do the same job I hired them to do.* In our view, titles you give early on *do* matter because one day, you will be in a position to hire a C-level person who actually deserves that title, and what are you going to do if your twenty-year-old with six weeks of work experience is already occupying that title? And titles, even if they don't matter to you, do matter to outsiders. If your title-inflated employee is not ready for what outsiders assume they should be able to do based on their title, you do a disservice to your employee by setting them up for failure or embarrassment while also tarnishing the reputation of your company in the process. So give titles that make sense for the experience level and responsibilities of your employee—you can always give them promotions as they merit them.

What do advisers do? What do directors do? Do I need them?

Advisers can do any number of things for your business, depending on your agreement with them. While not employees of the company, they might get stock options in exchange for some level of commitment (hourly or otherwise) to provide advice or other work (such as looking over your financials or making introductions to retailers).

Directors may also give you advice, but, more importantly, they are there to make sure you are doing what you are supposed to be doing to increase your shareholders' value. Basically, they are your supervisors and formal members of your corporation with fiduciary responsibility to make sure the company and *you* are doing as you should. This is the key difference between advisers and directors. While they are both there to keep you on course (or to course correct as needed), advisers are not accountable for your performance, but directors are (and directors have been known to fire founders and CEOs if they see the need).

As for whether you need them, if you have shareholders, you are required to have a board of directors and regular meetings so that the directors are informed and up to date on your company's performance. You do not *need* advisers, but we have certainly benefited from ours.

A Final Word

We can't help but give a plug for our sisters out there. We mentioned earlier that some of the people who might be willing to work for lower pay and longer hours were those with no experience or who were "unemployable." Don't overlook someone who doesn't have traditional experience—we're thinking of women who have been out of the workforce for a while or moms who have been raising families. They might not have work history, but that doesn't mean they don't have skills. Did they run the school auction? Organize a food drive for the church or temple? Rally people to the capitol to support legislation? Volunteer on a political campaign? Think of your network and who is a real go-getter and make-things-happen kind of gal—and see if she wants to come on board.

Chapter 9: Setbacks, Screwups, and Successes

It Must Be Tuesday

We've shared a lot of stories here. But for every tale of success, there are hundreds more tales of setbacks, screwups, and utter failures. What sets an InventHer apart from all the wannabe entrepreneurs is the willingness to forge ahead in the face of failure. Starting a business is never a smooth process. When you add the complexities of producing and selling a hard good as opposed to a service, the hurdles multiply. After talking to many InventHers, we discovered a few common mistakes as well as ways to get past them. Let's break them down.

It Always Costs More Than You Think

Anyone who has embarked on a major home-improvement project probably knows this excellent rule of thumb: budget for at least 25 percent more than the contractor tells you it will cost. Plans usually only take into account the best-case scenario—where no wood rot is discovered, the foundation doesn't need earthquake retrofitting, and there is no DIY electrical work to remedy. Rare is the house project that doesn't come with surprises. One of the InventHers we mentioned earlier referred to having the "shit happens" fund. We think this is a *great* idea and should be a key part of any business plan. No matter how you are raising funds, whether by bootstrapping, crowdfunding, or tapping your rich uncle, cutting corners is not going to get you where you need to go. Plan on having an emergency stash for the inevitable unforeseen costs.

It Always Takes Longer Than You Think

Again, we'll fall back on our home-project analogy. Do you know any major project that ever finished on time? Neither do we. When you're full of enthusiasm and eager to get started, you want things to go fast, from market research to prototyping to production. The journey from idea to product is a long one, and each step has potential for delays. You'll find yourself waiting for responses to critical emails or phone calls, waiting for prototypes, waiting for prices and proposals, waiting and waiting and waiting. Look at all the cautionary tales from Kickstarter or Indiegogo—projects that were promised to ship to backers within months instead dragged on for years. We believe most of these project owners believed they could deliver on time, but they had no idea of the amount of work between a campaign and a finished product.

It's harder to mitigate this one, but if you're thinking of quitting your current job to pursue your idea, think about an intermediate strategy. Can you keep your day job while also getting the idea off the ground? Is there another way to keep money coming in while you are spending money developing the product? Don't commit to any dates for final products until you are very, very far along in the process.

It Always Comes Down to People

Setbacks and screwups often revolve around people—either trusting the wrong people or failing to bring in the right people at the right time. Those InventHers that had a good network or were lucky enough to have a cofounder who was a loyal spouse or partner had a far easier time than those who relied on new business partners or investors who weren't on the same page. Some business partners or investors stole ideas and took them to market themselves. Some just wasted the time and money of InventHers, making promises they didn't keep or offering bad advice. The bottom line is, few InventHers are truly one-woman shows. For every successful woman, there are people behind her helping her succeed, from family to employees to investors to customers. You can't do it alone. Choose wisely who is along for this ride. Protect yourself as best you can from unscrupulous players. When it comes to employees, we will state the rule again: "Hire slow, fire fast." In the same way, take the time to be sure of relationships before entering into contracts or agreements, whether with a factory or a PR agency. These people can make or break your business.

So how can you stay on track in the face of so many obstacles? Mistakes are going to happen. One thing we suggest is to draft a clear and simple personal mission (why your company/product exists) or vision (what you will achieve) statement, and your objectives. This seems like a simple exercise, but eventually you will hone and add to it to create a document complete with strategic pillars and drivers of growth. For now, even your initial statement can serve as a touchstone to refer to when things get murky. It can be as basic as bullet points:

- create the best helmets made from potato peelings

- highlight the plight of potato skins

- market to the cycling community first, then move to football and motorcycle helmets

- create a work environment where people come before profits

- only work with people dedicated to the same cause and who will be like family

When wrestling with a decision, referring to some original guidelines can be helpful. As a founder of a brand, you're going to have to make thousands of small and big decisions every day. Should you take on this investor? Should you spend the money to be at this trade show? Should you fly to New York for a media event or should you drive? There are so many good ideas, and you'll be pulled in many directions. We can't stress how often founders have heard, "You won't want to miss out on this opportunity! You need to sponsor this event for $5,000 or you'll miss the boat." Or "Hire a marketing director now, even though you're not quite ready. She won't be available in another month." Or "I know you're planning on expanding into this other market eventually, and this is a great way to get the first foot in the door." When in doubt, go back to the plan. It can even include your personal guardrails, such as "I won't miss my kids' birthdays / my anniversary / my annual girls' weekend with college friends." Readjust when necessary. But keep your eye on the prize and the big picture.

The Ultimate Screwup (or Insurmountable Setback)

OK, so you did all that and still screwed up. You had the choice of going on *Shark Tank* or delivering the keynote at the annual gathering of the Buy Nothing movement, and you chose the latter. Welp. Time to put it behind you. To paraphrase Scarlett O'Hara, tomorrow is another day. Learn from it and move on. And what if you have the ultimate failure: your product has totally fizzled, your crowdfunding project hasn't been funded (or worse, you couldn't deliver and you have to refund money you don't have), or you built the most awesome product ever and produced 20,000 units only to discover a nearly identical product beat you to the market by two months and has a significantly lower price? This is "know when to fold 'em" time.

Closing up shop is a significant moment. It's not something you get over in a week as you move on to the next thing. Allow yourself to mourn—this is a great time to get out a few pints of ice cream and put on some Julia Roberts films. Have a pity party for one, then take a deep breath and begin disassembling what you've built. You may have to let employees go or—gulp—let retailers or customers know you're not going to be able to fulfill orders. This is just business, and doesn't reflect your worth as a human or even as an InventHer. This isn't failure, just business reality. There is no point in going further with time, money, or energy if the handwriting is on the wall. Proceed with dignity and transparency.

First, we suggest a thorough and honest postmortem, and a little more marketing. This time, the marketing messaging is about you. Look at what made you come to this tough decision—was it a single factor or many? Was it the factory you chose that screwed you over and lied about what they were capable of producing? Was it the sudden spike in aluminum costs coupled with new tariffs? Was it that employee you just discovered was embezzling money? Was it your poor online sales with no other channels? Or a combination of all of the above? Get honest with yourself, and then get your story straight for the world. Because you're going to be back. You're not back to square one. You've learned a ton from this journey. Cry alone at home with your mint chocolate chip, then put on your best face and go out to your investors, customers, the press, or whoever is asking, and give them the message you've prepared: "Unfortunately, PotatoSkinz Helmets is closing as of June 1. While we're confident the future of helmets is tuber based, current market conditions are such that it doesn't make sense for us to continue retail operations at this point. We will be regrouping to better serve this vital need. We thank our staff, investors, and friends who supported us in our root vegetable–based dream."

 HOT TIP: If you fold, no apologies. It's just business.

While this is your public statement, make sure you document for yourself where things went sideways. Your postmortem is not just in your head—get it on paper or a spreadsheet. Was it the factory? Next time, build in a redundancy. Employee embezzling? Next time, build in some more oversight. Raw materials costs? Look at alternative materials next time. Poor online sales? Review your marketing plan and budget, and see where you could have filled in some holes. Brutal honesty with yourself is part of this journey.

But the journey must end at some point—every extra minute you spend prolonging the inevitable is a minute you're not spending on your next big idea. Every dollar you spend on a doomed project is one that can't go to your next invention. Take this experience, learn from it, and you'll be able to ramp up your next idea even quicker. Open that pinot grigio (we're there for you) and start sketching out a new plan.

InventHers on Their Biggest Mistakes

SARAH BLANKINSHIP, SIVA PATCH

Biggest mistake: Not believing I could do this. I didn't even start out as CEO at my own company. I had an early partner who violated my "no asshole" policy. I finally ditched them and elevated myself to where I should have been in the first place.

CHEZ BRUNGRABER, GOBI GEAR

Biggest mistake: Hiring other people to do my work for me at the beginning. I handed over the reins. I hired a PR firm, but I didn't learn anything. When you're starting out, it's easy to hand over, but it's better to collaborate so that you learn things yourself.

FRAN DUNAWAY, TOMBOYX

Biggest mistake: Committing too early in a rush to find early investors. It's important that when building that founding team, you work together for about six months before committing to any partnership terms.

JUDY EDWARDS, SQUATTY POTTY

Biggest mistake: Thinking someone knew more than we did. It's easy to forget to follow your own gut and to think someone would be your savior. We hired someone to make a video for us—she was supposedly an "expert." The video was a total flop. Then we found out she'd already filed for a design patent on her own stool. We had to spend a lot of money on shutting down copycats.

LISA FETTERMAN, NOMIKU

Biggest mistake: Trusting an untrustworthy investor. They ran off with my ideas to a competitor. Protect your intellectual property!

MICHELE MEHL, EXCY

Biggest mistake: Quitting my job too soon. I knew the early stage would take a long time, but I didn't really fully grasp how long. I wanted to go fast—go big or go home, zero to hero. I cut out of my old company too quickly. Then I had to ramp it back up to pay the bills. I wish I would have kept it going longer, since I was bootstrapping.

SIRENA ROLFE, TEMPUS HOOD

Biggest mistake: I spent too much time doubting myself. Lots of times I thought, *Is everyone going to think this is as great as I do?* They don't have to. Just enough people need to. I also took too much advice from too many sources. If my boyfriend's friend said something, I'd listen. If people with more experience gave advice, I'd do it. I had to learn to trust my gut.

NANCIE WESTON, GRAYL

Biggest mistake: Not hiring people who had the same passion and vision as I did about purifying water and reducing plastic waste. I hired some folks who were in it only for the money—it was not a passion for them.

Wait, What If Things Go Right?

We've talked a lot about setbacks and hurdles, obstacles and problems. Here's a new one: What if everything actually comes together? You've tackled and solved each problem, your little product is selling, and your business is starting to grow. Sales are increasing exponentially and work is piling up. This is a good problem to have! It's also time to scale.

We've already discussed scaling your business. It's not simply adjusting as you get bigger. It's handling growth while gaining efficiencies. The costs you had to rack up for each sale at the beginning should not be the same now. Maybe you had one guy who packed all your boxes and worked twenty hours a week. Now that same guy has to work fifty hours, and you're paying overtime. This isn't efficient. Maybe you did all your own bookkeeping or managed the social media accounts yourself as you built the business. It's time to reassess a lot of things before you become a victim of your own success.

EXPANDING YOUR TEAM

The first thing to do as you scale is acknowledge you're no longer a fledgling business. You've flown the nest and you've got different needs. Part of the excitement of running a business is that it is never the same—what you did six months ago is ancient history. As a business owner, you're reinventing yourself (at least) annually. We've already talked about how this impacts your team—those fabulous generalists you had who could do everything need to now become specialists. Your right-hand woman who did all your marketing, packed boxes for fulfillment, paid your quarterly taxes, and did some child care for you on weekends cannot continue in this role. It's a recipe for unhappiness all around. Find a niche for her—because now you need a full-time marketer, a full-time operations person, a full-time accountant, and a part-time Mary Poppins. This is part of your business evolution and growing pains are common.

EXPANDING YOUR OFFERINGS

So you're sipping champagne (you've moved up from pinot grigio) because your product is a success. But the questions start coming: What's next? It's rare to have that one product that never changes or evolves but continues to sell. The stone wheel was a hit back in caveman times, but we now need something a little smoother for our SUVs and bicycles. Your investors and the press are going to start asking for some future product plans. It's now time to put together a product road map.

Basically, a product road map is a high-level overview of the direction of your product offerings. Think of it as a strategy document. It is usually developed by a product team, and lists a timetable for future product development and release. This will include iterations of your current product—new colors, materials, sizes, features, and add-ons. But it could also include new complementary product offerings. You've killed it with the new airplane pillow, and people know your brand as the go-to one for travel products. How about offering matching travel socks for in-flight comfort? Did the fancy pet collar take off? Matching leashes and pet beds may be in your future.

The bottom line: Invention is continuous. So is creating a successful business. You'll reinvent your business many times after your initial successes. Soon, it will be habit to always be thinking twelve, twenty-four, thirty-six months down the road. InventHers are always looking ahead!

Gloria Hwang
Thousand

THE CEO IS THE GLUE THAT BINDS

Gloria Hwang didn't wear a bike helmet. She wasn't a serious cyclist; she was a casual, recreational rider, biking to the local coffee shop or along the beach bike path. She cared about style and convenience—the ability to hop on and off a bike without looking dorky or having to carry around a helmet once she got to her destination.

That all changed when she lost a dear friend and mentor in a bike accident. He died from head injuries sustained in the crash. She decided there had to be a better way. She also believed people would be more connected to their communities and would get out of their cars and bike more if there was a better alternative to the helmets on the market.

Gloria worked at Toms shoes, and she knew something about taking a product to market. She sketched out a business plan, and nine months later, she had a prototype of her helmet: a stylish, retro-vibed beauty that didn't look anything like the ventilated bike racer helmets that dominated the market. And it had a clever feature—a "Poplock," a hidden hole where a bike lock could go through so the helmet could stay with the bike.

She decided to launch a Kickstarter campaign to raise some funds for manufacturing, and to test the interest in the market. She set a goal of raising $20,000 to get started. Her projections were wildly off—she raised the $20,000 on the first day, and close to $250,000 by the end of the campaign. This was far past the number she'd identified as allowing her to quit her day job and pursue this dream. Now she had money to produce and knew there was demand for a nontraditional stylish helmet.

While Gloria was able to quit her day job, she had many other jobs in store for her as CEO and founder of her new company. No job is too small for a CEO, and she jumped from task to task. She found her manufacturer, but struggled through production, as no one had seen a helmet like hers before, and they didn't understand the Poplock or her need for meticulous attention to detail. She ended up going to China and sleeping at the factory for three months while she walked them through the specifications. But her manufacturing worries weren't over. Her first big batch of helmets, all 2,000 of them, had a small cosmetic flaw: the glue the manufacturer used on one part didn't take properly. (Nothing that impacted safety or fit—just looks!) Thus began a month of hand gluing 2,000 helmets in her garage. She enlisted friends, with sometimes thirty coming over at a time to glue on this small part and make it just as she wanted it. She went back to China and made sure the second batch had a better glue and was happy with the results at the factory. And then . . . it turned out that the second glue didn't do well in the heat of the shipping container on its ocean voyage and melted a bit. This time, however, some of the helmets were getting shipped directly to customers, so she couldn't glue them in her garage. One order for 400 helmets was being shipped directly to a customer in Amsterdam. Gloria jumped into action as CEO and Chief Helmet Glue Technician. She hopped a plane to Amsterdam, intercepted the shipment, and spent the day hand gluing the same part to all 400 helmets in her hotel room, making sure the customers would be getting exactly the quality helmet she'd promised.

Gloria's tenacity—or dare we say, "stick-to-itiveness"—is what has made her a successful InventHer. This was her path for funding, as she chose to bootstrap her way into seven-figure revenues. This was her path for marketing, as she individually reached out to reporters and got great press coverage that boosted sales. She also worked at bike events, selling helmets from a tent and boosting brand awareness as thousands of cyclists rode by seeing the name "Thousand" on helmets. Today, the brand is thriving, as Gloria and her growing team are thriving, because Gloria leads by example, jumping in to address every challenge.

Gloria had doubts, like many InventHers, but ultimately succeeded by putting trust in herself to do whatever it took to make Thousand successful. "The thing I did right was not listening to the experts," says Gloria. "Because experts will tell you what will be successful based on their previous experience. I knew that there was a new market for helmets about style and convenience, because that market was me."

Sarah Blankinship
Siva Patch

OVERCOMING THE UNCERTAINTIES OF AN EMERGENT INDUSTRY

Occasionally, roadblocks are built in to your industry but you don't discover them until you are well down the path of product development. Sarah Blankinship wasn't looking to innovate in the wellness and health-care space. Her mother's sudden stroke forced her to look at new pain-relief methods for a loved one. This resulted in the creation of the Siva patch—a transdermal cannabis patch.

Sarah was a veteran of the tech industry, where she loved using data to make business decisions. But when her mom had a stroke and was released from the hospital in constant pain and with a purse chock-full of opioids with scary side effects, Sarah decided to explore more natural remedies with the support of her mother's eldercare doctor. Cannabis solutions were becoming more mainstream with legalization for medical uses, but the industry was woefully behind. Sarah found one company that was making patches, which she thought would be the ideal delivery method for her mother, but was appalled at their production standards and facilities. "I went to visit their facilities and was just horrified. The lab was filthy; they were set up in a storage unit, not a sterile lab environment. There were materials in dirty buckets, no adherence to good manufacturing practices," she said. "I immediately started making calls to buy my own equipment and start a proper lab."

Sarah spent a year and a half working out of a lab in Ohio, talking to chemists and patients, researching all the different uses of and the market for cannabis wellness products. With experts citing efficacy for uses from chronic pain to hangovers, she knew she had a market. Her mission: to patch the planet. She tested and tested and finally came up with the Siva patch.

It turns out that creating the patch wasn't the biggest challenge. Sarah found herself in a regulatory no-man's-land where nothing was clear—a gray area called nutraceuticals. Every day was a new revelation: she had to get certified by the DEA to work in the lab, she needed FDA approval, she discovered cannabis was still a Schedule 1 drug, like heroin or methamphetamine. Illegal on a federal level, legal on a state level. She could use or market the word *hemp* but not the words *marijuana* or *cannabis* or *CBD (cannabidiol)*. An FDA adviser had to ensure she wasn't making medical claims. In short, she had to tell the story and tout the benefits of her product without crossing certain boundaries.

"We have never run an ad. We want to be on nobody's radar, which makes it hard to market," admits Sarah. "We're going to have to grow slower than we want, but that's OK. We're being patient and playing a long game."

For the moment, Sarah focuses on building a business model that will scale, within the confines of all the regulatory challenges she faces. She's happy to work both with partners to white label the patch (a process that lets other companies brand it as their own) and with wholesalers. She has pinpointed her market as active people with injuries and successful aging people. She's attended conferences on the global health crisis of stress, and is looking at clinical trials. Her marketing strategy relies heavily on giving out samples and getting repeat buyers on Siva's website. So far, it's working. Sales were up significantly in 2018.

As she navigates the evolving regulatory landscape of her industry, one measure of success is clear. Her mom is doing great. "She is awesome. She's my best patch consumer," says Sarah. "She has great ideas and has helped us understand issues about accessibility."

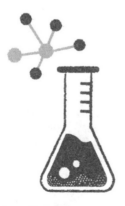

Chapter 10: Work and Personal-Life Balance

Hey, What a Pipe Dream

When we started thinking about writing this book, aiming to empower women to bring their ideas to life, we knew we had to tackle the topic of the work/life balance myth. Yep, myth. Work/life balance seems to imply there is some perfect balance point, a magical percentage of time or ratio of effort where the scales will balance, the stars will align, your business will thrive, and your personal life will be utterly fulfilled. When you find this balance point, you rake in millions, your spouse dotes on you or your dating life is robust and exciting, your home is sparkling clean, and your children are happy little students of the month who always offer to do the dishes before tackling their homework. In this perfect life, there are no science-fair projects. You run your business between the hours of nine and five, then attend yoga classes three times a week and still have time for your book club and maybe a gourmet cooking class. You meet with investors for cocktails, then head out for a night on the town. Weekends are for fun with family and friends and healthy self-care, so you're ready for Monday when you head back to the office. When you find the balance, the angels will sing.

Kiss that fantasy goodbye. There is no perfect balance. In fact, the new term you hear these days is *work/life integration*. Digital access has blurred all the old lines between work and home, and most people, in every field, have difficulty disconnecting from work when our phones and laptops are never really out of sight. For entrepreneurs just trying to get a business off the ground, it's worse. For female entrepreneurs, we're going to go out on a limb and say it's even harder. Studies have shown that women are still doing the lion's share of many domestic tasks. We're dealing with kids, aging parents, spouses who travel, pets who need attention, homes that need upkeep. Just because you've invented the World's Greatest Whatever and you are working eighteen-hour days doesn't mean that your cat doesn't need a flea dip or that it's not your turn to provide gluten-free, allergen-free, vegan homemade snacks for the soccer game.

We want to give you a real, honest assessment of what it is going to take to bring your product to market. Like any baby, your fledgling business is going to need your attention every day, and it won't disappear for evenings and weekends. Being an entrepreneur, like being a parent, is an all-consuming endeavor, but like parenting, it's also worth it. Every long hour you put in isn't for some faceless corporate employer—it's for *you*. It's for your and your family's future—what could be more worthwhile?

Make a Realistic Plan

You know you'll need a plan for your business endeavor—a road map of what you need to do to get this product in consumers' hands and money in your pocket. Make a prototype, create a marketing plan, find a manufacturer, get some funding, and so on. Make a life plan as well. This doesn't have to be a PowerPoint—you can sketch it out. Look at the other obligations in your life: parenting, pet care, volunteer gigs, exercise regimens. It's time for a realistic assessment of what you can do, what you need to give up, and what you can outsource. While two hours a day for your yoga practice or spin class (with travel time) may seem nonnegotiable, we're here to tell you that ain't gonna happen. You're going to be working more than you think. It's great that you've been a super active alumna for your university and have spearheaded the annual fund-raiser or coordinated the local interview efforts for applicants for ten years—but this is the time to sit that one out. Women in particular want to say yes to everything. We get it. Saying no is hard, but it's part of being an entrepreneur. You need to be able to ruthlessly prioritize, as one of our InventHers stated. Prioritize, delegate, and have a laser-like focus on the business.

The first part of making a plan is having a realistic view of all your current commitments. It's easy to think, "Well, sure, I run the church coffee service every Sunday, but I don't count that, since it doesn't take much time." Be brutal—it's not just time, it's the energy and headspace you can't afford. Think of how much time you spend emailing or making phone calls for these small activities. If someone is sick, do you have to scramble to find a replacement? Do you have a needy volunteer you have to manage with frequent calls and emails? This would be the time to step back and pass off coffee-service duties to someone else. Say a polite no to new volunteer opportunities or social obligations that are going to cause anxiety later. Pro tip: If your best friend wants to know if you want to see that hot new show and she can get tickets for a date four months from now, politely decline. You have no idea what you'll be doing four months from now and what will be happening that day. If there is a crisis then or you need to travel on urgent business, you'll just have added stress if you have to bail on your friend at the last minute.

Delegation is the next thing. We can't get out of all our obligations. Those kids need to eat and the dog needs walking. Take a look at what you can delegate or outsource. Child care is the biggest, of course. Don't get involved in the Mommy Wars, trying to live up to some unrealistic expectation of doing it all. We support *all* women. Moms who want to stay home with their kids? Great! Some of our InventHers did just that for years before venturing into this world. (Hilary left her law practice to raise her kids full-time when they were young.) Michelle Obama perhaps said it best: "I tell women, that whole 'you can have it all'—nope, not at the same time; that's a lie. It's not always enough to lean in because that shit doesn't work." So be realistic. Don't set up a recipe for stress and unhappiness at both work and home. If you're willing to be a working mom *and* an InventHer, be fine with this decision. You are responsible for having your kids safe, fed, and cared for. You don't need to be there to serve every juice box and do every school pickup. A spouse, family member, or nanny can do these things as well. Paying for child care will allow you to focus on your business when you're apart from them, so when you come home you can have quality time with the kiddos. (And this should go without saying, but support other women who may choose differently than you—we all have our own journey and aren't in a position to judge anyone else's.)

Outside of kids, look at other obligations. Look into a dog-walking service (or hire a responsible neighbor kid) to make sure Barkimedes gets some exercise. Yard work? Another thing to outsource. Keeping that house clean and laundry done? This is perhaps the easiest thing to take off your plate. In our opinion, aside from child care, outsourcing housekeeping is the number-one thing that can free up time for your business and also help keep you sane. No one wants to come home from a twelve-hour day to a house that needs two hours of work before you can make dinner. That brings us to meal prep. Budget and plan for the easiest, least time-consuming meals you can. This can mean pizza deliveries, Costco chickens, or Sunday bulk meal prep for the week. Every family has a different solution. If you love to cook, great. But like that spin class, two hours a day of meal prep and cooking aren't going to be sustainable. Make a plan that works for you.

 HOT TIP: Housekeeping is the easiest thing to outsource.

Set Some Boundaries

Part of your plan must include some guardrails you don't cross, the sacrosanct rules of your life that are nonnegotiable. It could be no work talk after ten o'clock at night. It might be a monthly date night with a spouse, or a bedtime story with your kids. Not traveling on birthdays or anniversaries. Attending religious services. Getting in a weekly run. These boundaries should be reserved for the highest-priority things in your personal life—the things that sustain you, recharge you, and nourish you. There's a big difference between time commitments that keep your marriage going and time commitments that keep your book club happy. You might take pleasure in both, but one is critical; the other is not.

Here's a tough one: boundaries can also apply to the length of time you're willing to commit to trying to get your product going—the amount of time or money you're willing to devote to your new business. As we noted in the previous chapter, it's always going to take longer than you think. That said, you might want to set out some basic guidelines, depending on your circumstances and family situation. For example, you could agree that you won't quit your day job until a certain benchmark is met, or that you will give your company three years before you call it a day and redirect energy to another project. Or you might have financial guardrails. For example, you're willing to put in $50,000 (or $200,000, whatever is appropriate) of your own money, but you're not going to empty your 401(k) or take out a second mortgage on your home.

We can't tell you what your boundaries should be. Everyone's situation is unique. Every product has its own gestation period, its own financial costs, its own daily and weekly time commitments. What we can tell you is that it is wise to think about your core values and how far you are willing to go for your business. Some might say, "You should give everything—your last minute and your last dime." We disagree. It takes a lot to be an InventHer—but it shouldn't take everything. What is the point of a bestselling product if you are not happy with the rest of your life?

Communicate!

You've got your plan, so now it's time to share it with the inner circle. By inner circle, we mean those nearest and dearest to you: immediate family members and partners, business and otherwise. First are those people you share a home with—spouses, partners, kids, parents. Everyone who sees you on a day-to-day basis should be aware of what you're about to embark on. This isn't just "I have a new job!"—which might be of passing interest to your kids or spouse—this is a major lifestyle change; and everyone who has expectations of you should now reset them. Most importantly, you should get some buy-in and support from this critical family network. They will be the people most impacted by your decision and should have fair warning.

Spouses, partners, and significant others take a special kind of communication, especially if you are financially interdependent. It's going to be a difficult road if you're going to take a financial hit (like quitting your job or digging into savings) getting your business launched and your spouse is not on board. Talk it out, and do it early on. Make sure you're hearing their concerns, and discuss financial boundaries. You might be surprised to find out that you have different comfort levels with risk. The best way to head off trouble is to talk through what you're both thinking. Address concerns ahead of time. While it isn't always possible to avoid all conflict, if you know that your spouse's deal-breaker is gutting the 401(k) or living without health insurance, you can factor that in. If you know that Sunday night dinner with the family is a priority, you can make a mental note as you plan your week.

You also need to communicate to your business partners (if any) and other key people so they know what to expect. Your cofounder should know that you turn off your phone at eight o'clock at night or that you won't be giving up your day job until you've pulled in a certain amount of revenue. They should know you fly once a month to spend three days with an aging parent and that you have an Ultimate Frisbee game every Saturday that you absolutely cannot miss.

InventHers on Staying Sane* While Bringing Your Product to Market

SARAH BLANKINSHIP, SIVA PATCH

Advice: Know how to ruthlessly prioritize. Ask yourself, "What is the one thing I must get done today?" Then do it.

Relies on: The friend network. My women's network is like therapy. We can have a glass of wine and laugh and cry.

CHEZ BRUNGRABER, GOBI GEAR

Advice: People want to connect with your story, and with you. So make it a good one. A good story makes it easier for people to connect. There are too many faceless corporations out there; people want something real and personable.

Relies on: Help. For the longest time, I had no problem. Now I have two little kids, and it's a lot harder. When they are home, I want to be with them. I hired an employee who keeps the company running while I explore an exit. Money well spent!

*Sanity is relative

AMY BUCKALTER, PULSE

Advice: Be mindful of your time. I have not had a balance I would recommend to anyone. In my entire career I have had work/life integration biased to the working side. I've been driven by the intellectual challenge, the drive to create the extraordinary. Most of my career hasn't felt like work because I felt passionate about it. I dove into Pulse after a breakup, and it was great to have something to focus on.

Relies on: Morning spin class and evening martini.

FRAN DUNAWAY, TOMBOYX

Advice: Don't expect balance. I believe that mediocrity is the fulcrum of work/life balance. There is no line for me and my wife (and cofounder) between work and life. It's all blurred—we love our lives, our company, our business. When we get home, we can say no more work talk after ten o'clock at night. But then we have to do it. We've made that commitment to build a brand for people who want to be heard.

Relies on: Laughter and communication. My wife and I are both good fighters and can express how we feel. We can deal with conflict and still make each other laugh.

JUDY EDWARDS, SQUATTY POTTY

Advice: Set your priorities. I was sixty when I had this idea. I had seven children. If this idea had come to me when I was thirty, I couldn't have done it. I had the time to spend on this. The doors were open. That first year was grueling. It consumed our lives. We had to start from zero. In fact, I was looking for something to do. I worked with my husband and son on this. It was a family affair. The timing was right.

Relies on: Faith and family.

LISA FETTERMAN, NOMIKU

Advice: No work talk in the bedroom. With my husband as a cofounder, we have a hard-and-fast rule. That's our safe place. We once had a really bad fight about work—like all night, from six at night to six in the morning. That's when we came up with the rule. We have never had another fight like that.

Relies on: A support network of nannies and a mother-in-law. Also planned date nights when they can watch the kids. If we can go two hours without talking about work, I consider it a triumph.

GLORIA HWANG, THOUSAND

Advice: I fluctuate on this, but mostly, there is no balance because if you're trying to accomplish something big, you've got to work harder and smarter than the person next to you. Also, if what you're building in your work life is fulfilling, do you really need work/life balance? Which is another way of saying I work a lot.

Relies on: Beach bike rides to clear my head. I like to go from Playa del Rey to Redondo Beach, and hit the seafood shack there.

MICHELE MEHL, EXCY

Advice: Get your family's buy-in. Their support is everything. The impact on my twenty-year marriage is significant. I walked away from a good salary. You're working really hard, but for no paycheck, and the house is in disarray. Then you think, *Oh, I should keep up the house more, because I'm not contributing financially*, but you can't.

I also asked my son's permission to do this start-up. It was one of the best things I ever did. I asked him because as I have seen during my times working with entrepreneurs, it was going to be a wild ride. It truly is a family affair. I just booked four flights and I will be gone for a good while, so I need them to understand. I feel like I set the stage for this when I asked.

Relies on: Scheduled date nights and continual buy-in and communication with my partner.

SIRENA ROLFE, TEMPUS HOOD

Advice: Surround yourself with a network of people who support you. This helps you get past your doubts.

Relies on: The thought that this is going to let me take care of my family. I can tell my mom she won't have to worry about not being able to pay an important bill. Also, the occasional vodka soda.

NANCIE WESTON, GRAYL

Advice: Dating is hard. Find men (or women) who truly honor and want to support you in starting your business. Some men say they like strong women, but when it comes right down to it, they really don't. They feel threatened, and egos get in the way. It's lonely running a start-up, but I love it.

Relies on: My friends. They knew I didn't have any money running a start-up and that I couldn't afford to go out to dinner, so I would have BYO parties at my house and ask everyone to bring their drink of choice and an appetizer. Your friends are your lifeline to keep you going.

Leslie Pierson
GoodHangups

A FAMILY AFFAIR

"I've never been a nine-to-five person," says Leslie Pierson, the inventor of GoodHangups, removable magnetic stickers that allow you to hang up artwork, posters, recipes, or paper items. "So I can't really give advice on work/life balance. When I have an idea, I want to *make out* with that idea. I'll stay up all night just thinking about it."

Pierson was definitely cut out for the entrepreneur life, and sees the benefits of getting her family involved. She had a previous business making decals for laptops. It was humming along, but was not particularly inspiring. The birth of her son in 2010 spurred her to start thinking about a new business, something that would be more innovative. In 2015, when he started bringing home tons of artwork from school, she had the idea for magnetic stickers and magnets that would allow her to easily change out last week's masterpieces for the latest creations. GoodHangups was born.

Pierson started with a Kickstarter campaign, a nerve-racking proposition as she had to be in the video—an uncomfortable feeling for someone more comfortable in the background. But she was game to push herself. The campaign set a goal of $5,000 but made $28,000, which was encouraging but not overwhelming. This allowed her to produce and ship her orders.

With zero marketing budget or expertise, Pierson tried to get her product into big-box stores. Often it was as simple as filling out an online form for consideration. "I figured, I can drink wine and do that." During one night of application blitzing, she filled out what she thought was an application to be featured on QVC, the home shopping network. It was actually a contest for products to appear on the *Today* show, and the prize was to be featured on the QVC network. A few weeks later she was notified that her product was a finalist, and she was off to New York to appear on camera again, this time in front of a TV audience of millions. She won, and after her QVC appearance, she sold $300,000 worth of product in six minutes. After that, an invitation to appear on *Shark Tank* followed, resulting in more success and a new business partner, *Shark Tank* host Lori Greiner, who is helping Pierson get GoodHangups into retail stores like Walmart and Bed, Bath & Beyond. Not bad for someone nervous about being in front of a camera.

Pierson had to make many choices along this journey to success, and credits the support and advice from friends who helped her along the way. In many ways, her personal network offered moral, if not financial, support. Her friends encouraged her to do the Kickstarter campaign and calmed anxieties about television appearances. But one touchstone through it all was her son. She kept him involved in the entire process, even having him appear in the Kickstarter video she made.

"I was talking with a female friend about how as we get older, we have more fear of failure. But we need to show our kids how to fail. When I thought about doing the Kickstarter or going on television for the *Today* show or *Shark Tank*, you get the chance to fail very publicly," says Pierson. "If it didn't work out, I could show my son how I failed, then got up and moved on. So either way, it was a win."

Unfortunately, Pierson is going to have to keep looking for a failure for her son to learn from, as GoodHangups has been all success. But that's a good lesson too. In fact, when her son was seven, he announced that he would like to create a card game called Taco vs. Burrito and sell it on Kickstarter. Leslie encouraged his creativity, and over six months of mother/son game time, they developed it. "We played the game every Saturday and Sunday, after we walked the dog. Each time we'd tweak it just a bit." Cute, right? She helped him make his own Kickstarter video and campaign, and their family lark raised $24,000. After shipping the games out to supporters, they sold the rest on Amazon, netting $49,000. The real reward, though, is the time that Pierson and her son got to spend together, and the entrepreneurial spirit that she has fostered in him. There is no way to put a price on that.

Conclusion

Why Not You?

Well, we've made it to the end together. It's time for one last huddle in the locker room before you go out and hit the field and score big. You're already ahead—you've gotten this far, which means you're serious about your idea and have an inkling that you can do this. Guess what? *You can.*

Don't get overwhelmed. It may seem like a daunting task, but we want you to think of all the great InventHers who came before you. Think of the many women, a few of them featured in this book, who had an idea for a product and made it happen. Women are notorious for underestimating themselves, and doubting minds start coming up with reasons why those other women were successful and why your idea or abilities won't measure up. To that, we have three words for you: *Why not you?* Why shouldn't it be you that has the next breakthrough product? Fortune favors the bold. Someone is going to invent that thing you're thinking about: Why not you?

Let's go back to the beginning and review why we wrote this book. There are other books out there for those looking to invent something, so why focus on women? Because we believe women are unbelievable problem solvers, creative thinkers, savvy networkers, and persuasive storytellers—all the things necessary to invent and take a product to market. However, we have been underrepresented, undersupported, and underfinanced. In addition to work outside of the family, we've shouldered and continue to shoulder a disproportionate share of family tasks, from raising children to caring for parents to maintaining a household. We are still working as hard, if not harder, than our male counterparts to earn a little over eighty cents on the dollar. (The statistics are even worse for women of color.) Female founders of start-ups got just 2.2 percent of venture capital in 2018.[16] Two point two percent! In 2018! Furthermore, sexism and misogyny are still deeply ingrained in many industries and, sadly, our society. We can't believe some of the stories we hear from InventHers and our network, from women being asked to bring on a male business partner to be taken seriously, to experiencing blatant sexual harassment from investors and suppliers. It happens—we don't want to sweep this under the rug. Be ready, expect it, and know how you're going to deal with it. These are the unique hurdles an InventHer will face.

Wait, wasn't this supposed to be a pep talk? We are certainly here to be your motivational speakers, but we're also giving it to you straight. So how do you handle it? By being the badass woman that you are. By refusing to be discouraged or hampered by those who stand in your way. And there has never been a better time for women to step up. Women's business networks are growing fast—there are female founder groups popping up every month, and you can quickly tap into a ready-made network of sisters willing to help. Women-initiated funds are growing too, so the money is starting to flow toward the XX chromosome set. Industry groups are gaining in power and membership—both of us have been involved with Camber Outdoors, a group dedicated to advancing opportunities for women in the outdoor industry. Mina has been a finalist in their Pitchfest at Outdoor Retailer, the giant outdoor-industry trade show, and Hilary has met multiple clients at Camber networking events. In short, women are finally starting to break those barriers to success, and there is no better time than now to start your own business.

There are advantages to being a woman in a male-dominated industry, such as the natural partnerships and shared goals with other women in that industry. It's often (but not always) a built-in network for InventHers. Take advantage of it, and don't be afraid to reach out and look for help. On the flip side, we humbly ask you to help those other women who look to you. Support other women's projects, put in some good (but honest) reviews for women-created products, and support women-owned businesses. Create a female-friendly place to work, and when you hit it big, invest in women's venture capital funds. Wherever you go, leave places a little bit better than you found them, especially for other women. By helping each other, we help ourselves. Use the female entrepreneur network while also making it stronger for those who come after you.

Let's make more lemonade out of the lemons women often get in the business world. Why work tons of hours to earn less than male counterparts when you could be spending those hours adding to your own bottom line? Long hours are tough for anyone—but when you are building something for yourself, you'll be a lot more motivated and fulfilled. Starting your own business is the perfect antidote to toxic male-dominated workplaces. Not only do you have the best boss—yourself—you won't be arguing with HR about why you deserve a raise, and if you need to take time off for family reasons, you won't have to ask anyone's permission. We've gone over some of the trade-offs—security, stability, that steady 401(k) contribution—but don't underestimate the job satisfaction that comes with pursuing your own idea and creating your own product.

 HOT TIP: Don't let sexist jerks bother you, but have a plan to deal with them.

Ladies, it's time. You have a killer idea. Don't get bogged down in details that can overwhelm you. The small stuff isn't worth worrying about, and you will figure out the big stuff (or find someone who can). Lots of women before you have done it. Think of all the InventHers in this book, plus the women throughout history who brought you disposable diapers, the fire escape, the board game Monopoly, the computer algorithm, Kevlar, the modern refrigerator, and Spanx. They all had two things in common: an idea, and the will to see it through to reality. There is no reason you can't join this pantheon of creative ladies. We've given you a road map to success. All you need to do is get behind the wheel and start driving.

You got this.

The Hard Data

- There are 11.6 million women-owned businesses in United States.[17]

- These businesses have generated $1.7 trillion in sales.[18]

- Nine million people are employed by women-owned businesses.[19]

- One in five firms with revenue of $1 million or more is woman-owned.[20]

- Seventy percent of women with children under 18 were in the workforce in 2015.[21]

- Four percent of patent applications are by women only (not as part of mixed-gender team).[22]

Do You Have What It Takes to Be an InventHer?

These are some questions to ask yourself:

- **Why?** Why do you want to do this? Think carefully about this one. Are you looking for a new income stream? In the short term, this isn't a good fit. If you're looking to be the boss and call the shots for your own business while learning something new every day, this is for you.

- **Are you ready to take some risks?** Ask yourself if you're in a place to take a few chances, financially and professionally. Look at your network and decide if you've got support in place, be it family, friends, cofounders, or a good child care provider, to help you be successful while building a business.

- **What are your strengths and weaknesses?** Think about your personality. If you like control and you stress over every detail, you might quickly become overwhelmed. If you like predictability and security, don't quit your day job. If you love a challenge, know when you need help, can delegate like a champ, and are flexible in your thinking, you may be an InventHer.

- **Do you enjoy sleep?** Do you prioritize a full ten hours a night? Yes? Well, move along. Being an InventHer isn't for everyone. (Note: If you're a mom, raising babies and young kids is excellent training for launching a business. The sleep schedule will be familiar.)

- **What is your relationship with take-out food?** While not strictly necessary, a penchant for eating out of delivery food containers while reviewing spreadsheets is a definite plus.

- **Want to feel like a badass boss lady?** Come sit over here with us. We like you already.

Acknowledgments

First and foremost, we are grateful for all the InventHers featured in this book for sharing their triumphs and mistakes so that we could all learn from their experiences.

Second, we are grateful to Jennifer Worick, Sasquatch Books, and Penguin Random House for taking a chance on two new authors and for turning our goal of producing an easily digestible, girlfriend-to-girlfriend invention guide into reality.

MINA YOO: I'm thankful to the Heroclip team, who not only contributed content and expertise to the book but also had to deal with their CEO showing up at work haggard and cranky after staying up all night writing. Thanks also to Alia Ellison for keeping my family and household chugging along throughout the writing of this book (and throughout the last eight years, for that matter). Big thanks to my parents, who have supported everything I have done in my whole life and who now believe that the world will run out of paper from all the copies that will need to be printed of this book, due to astronomical demand. Finally, lots of love to my hilarious and amazing kids, Kai and Mila—who provided hugs, kisses, jokes, and massages as "resets" to writer's blocks—and to my husband, Mark, who deserves endless gratitude for not only being a sounding board but also my bartender, mixing up delicious cocktails to celebrate particularly productive writing sessions (which happened a lot). I love you!

HILARY MEYERSON: I'm so grateful to have a team that supports and sustains me every day, as I juggle running my business, writing a book, keeping the house stocked in cat food, and running off for a solo hike now and then. I want to give a shout to my fierce female network, without whom life would be much duller. My college ladies, whom I met in that dorm in Vermont so many years ago: Lynn, Susan, Holly, Lisbeth, and Stephanie. Our girls' weekends are one of my pillars. Thanks to Julie for the daily chats and texts that keep me going—I treasure our friendship! Thanks to Christine, who has been taking walks with me since kindergarten, even if they are now by phone. And I couldn't do it without happy hours with Meghan!

But most of all, I thank from the bottom of my heart, my tribe: my husband, Randy, for his unflagging support of me and my writing for nearly thirty years, and my incredible kids, Henry and Harper, who bring me joy every damn day. You guys keep me laughing and thinking and writing, and I am so, so grateful for our family. I love you guys so much.

Notes

Introduction

1. Jessica Milli et al., "The Gender Patenting Gap," Washington, DC: Institute for Women's Policy Research, July 21, 2016.

2. Bridget Brennan, "Top 10 Things Everyone Should Know About Women Consumers," Bloomberg.com, January 11, 2018.

Chapter 1: The Initial Idea

3. "FinalStraw: The World's First Collapsible, Reusable Straw," Kickstarter, www.kickstarter.com/projects/finalstraw/finalstraw-the-worlds-first-collapsible-reusable-s.

Chapter 2: Prototyping

4. "Regulations, Mandatory Standards, and Bans," United States Consumer Product Safety Commission, cpsc.gov.

Chapter 3: Customer Identification

5. Statista, "Travel Goods Retail Sales in the United States from 2000 to 2016 (in Billion U.S. Dollars)," www.statista.com/statistics/252760/travel-goods-retail-sales-in-the-united-states/.

6. Ben Midgley, "The Six Reasons the Fitness Industry Is Booming," *Forbes*, September 26, 2018.

7. Harris Williams & Co., "Physical Therapy Market Overview," February 2014.

Chapter 4: Distribution

8. Think with Google, "The Need for Mobile Speed: How Mobile Latency Impacts Publisher Revenue," September 2016.

9. Lauren Thomas and Lauren Hirsch, "Amazon Says This Prime Day Was Its Biggest Shopping Event Ever with 100 Million Products Sold," CNBC.com, July 18, 2018.

10. Gary Lee, "Why 90 Percent of Sales Still Happen in Brick and Mortar Stores," Retail Technology Review, October 17, 2017.

Chapter 5: Marketing

11. Aaron Smith and Monica Anderson, "Social Media Use in 2018," Pew Research Center, March 1, 2018.

12. Kurt Wagner, "Study Finds 77% of College Students Use Snapchat Daily," Mashable.com, February 24, 2014.

13. Irene Voisin, "Here's How People Shop on Pinterest," Pinterest for Business (blog), March 8, 2018. https://business.pinterest.com/en/blog/heres-how-people-shop-on-pinterest.

14. Russell Redman, "Meal Kit Players Adapt to Changing Market," SupermarketNews.com, November 30, 2018.

Chapter 8: Building Your Team

15. Haje Jan Kamps, "Breaking a Myth: Data Shows You Don't Actually Need a Co-Founder," Techcrunch.com, August 26, 2018.

Conclusion

16. Emma Hinchliff, "Funding for Female Founders Stalled at 2.2% of VC Dollars in 2018," Fortune, January 28, 2019.

17. American Express, "The 2017 State of Women-Owned Businesses Report," January 2017, 3.

18. American Express, 3.

19. American Express, 3.

20. American Express, 4.

21. Mark DeWolf, "12 Stats About Working Women," US Department of Labor (blog), March 1, 2018. https://blog.dol.gov/2017/03/01/12-stats-about-working-women.

22. Eileen McDermott, "USPTO Report: Only Four Percent of Patents Name Women-Only Inventors Over the Last Decade," IPWatchdog.com, February 13, 2019.

Resources

Repeat after us: The internet is your friend. You and Google can solve a lot of problems as you grow your business. In fact, your problem might be too much information. We've compiled a list of resources we've found helpful at various stages, whether you are a one-woman show or you've built a team of twenty to get your product manufactured and promoted. This list isn't exhaustive, but it can help you get started. We are not endorsing these websites or making any promises—just telling you what worked for us.

Market Research

Bureau of Labor Statistics (bls.gov): If you are trying to learn about the work circumstances of your potential customers or how the labor market is trending, check this out. It breaks down full-time and part-time workers by common demographic and geographic characteristics.

Google Forms: An alternative to SurveyMonkey, this is an easy-to-use way to get quick information from whoever is giving you feedback on your product or product idea. Then it magically flows all the responses into—where else?—Google Sheets.

Pew Research Center (PewResearch.org): Like the census data, this site is free. Instead of a government agency, Pew Research Center is a nonpartisan "fact tank that informs the public about the issues, attitudes, and trends shaping the world." They do lots of public opinion polling, demographic research, content analysis, and other data-driven social science research. Their research on social trends is particularly good.

SurveyMonkey (SurveyMonkey.com): It's an oldie but a goodie, and it has a free version. Create simple polls and gather some data on the cheap. There are also free poll features on Twitter and Facebook.

US Census (census.gov): Need to know the size of your potential market? How many women aged thirty-five to thirty-nine years old are single and with children under eighteen at home? How many female students are enrolled in colleges across the country? Look at your census data.

US Small Business Administration (sba.gov): You pay your taxes, so why not take advantage of the insights this government agency puts out? The site features a lot of market research information, including a neat little tool called Size Up that lets you compare your business to other small businesses. It also includes information on starting a business, and some funding options that you might not have known were available.

Content Creation Resources

Adobe Spark (SparkAdobe.com): This Adobe app can help you create social media content and other visual content in a snap. There is a free version (that includes the Adobe logos), but a minimal monthly fee is worth it for all the upgrades. Choose a template like "Instagram post" or "Facebook ad" or "teaser video," and get to work. It's simple to use, and you can drag and drop images, change text and fonts, etc. All in all, this is a great, robust tool.

Canva (Canva.com): A cloud-based graphic design tool, Canva offers a limited number of options to use for free. You'll have to pony up a nominal amount for some of the more useful features. It is one of the older tools in this category and can be a little clunky sometimes, but its strength is its simplicity.

Fiverr (Fiverr.com): This is an online marketplace for a wide variety of services. Their name came from their origin, where jobs for quick illustrations or other content started at five dollars. You could have someone create a logo, Photoshop your business name onto a cloud, or hand out flyers around a campus. Some services are out there, like people who will sing "Happy Birthday" to you while wearing a banana suit (and send you the video afterward, of course). Prices have gone up, but it's still a good place to look for some simple design work.

Social Media Management Tools

There are multiple tools out there to help you manage your social media accounts so you don't have to engage on each one individually. These tools help you schedule posts, interact with followers, measure success through detailed analytics, find influencers, gain followers, and more. There are many niche tools that do a very specific thing well—like finding relevant followers on Twitter (Tweepi) or curating and scheduling out "evergreen" (i.e., not particularly timely) content (Edgar). There are tools that just help you find social influencers (BuzzSumo, or Followerwonk). Determine what you need. At the beginning, you'll need to at least figure out what channels you want to be on, then manage and schedule content for everything on one dashboard. New tools are popping up monthly, and others are shuttering as rules change and the social media channels keep updating their APIs (application programming interface—how the developers access the channel) to protect their business models and protect consumer privacy. It's a fast-moving industry, so you'll need to stay on your toes. That said, we have a few suggestions based on what has worked for us.

A word about scheduling. It's a must. You might think you will have time to post organically every day or two to your social media outlets, but you won't. Scheduling makes it easy—also, you can post the same content to multiple channels at one time. A caution: you can never rely solely on scheduling. Part of the appeal of social media is the instantaneous feedback and reaction to real-time events. A savvy social media manager can capitalize on what is trending at the time. For example, during Super Bowl XLVII in 2013 there was a power outage that suspended the game. Some genius at Oreo came up with a quick graphic of a cookie and milk and tweeted, "Power out? No problem. You can still dunk in the dark." That tweet went beyond viral. No scheduling could have foreseen that. On the negative side, scheduled tweets can be a disaster if no one is paying attention. The NRA was lambasted when a tweet that said, "Good morning, shooters! Happy Friday! Weekend plans?" went out on the morning of a mass shooting tragedy. A major concert promoter of a Radiohead show also failed to cancel scheduled tweets after their concert venue collapsed, tragically killing someone. Their initially benign "Share your photos of tonight's show using the hashtag #RadioheadTO!" tweet—sent after a nonscheduled "Tonight's show has been canceled" tweet—did not go over well. So just remember that scheduling is handy, but you still have to monitor.

Hootsuite (Hootsuite.com): The granddaddy of social media management tools has been around since 2008. As with others, you pay by how many users have access. For a larger team, it can get pricey. But you can view multiple streams at once, and we find it's particularly great if you're managing multiple Twitter accounts. You can monitor certain hashtags, and of course post to all your social media feeds at once. This is a great place to start if you're new to social media.

Sprout Social (SproutSocial.com): This one has been a favorite of ours. Like Hootsuite, you can publish to multiple accounts and channels. You can delegate tasks to certain team members (like responding to a customer on Twitter), and see your full publishing calendar for the month in one view. We also really like the reports it generates—it's great to see your activity for a certain time period. It's a little more expensive than Hootsuite, but we like the interface better. Try them both and see what works for you.

Keyword Tools

Keyword search tools: There are a bunch, and many are free. Start searching for terms, or keywords, that you think people might use to find you. Is it *airplane pillow* or *inflatable travel pillow* or *travel neck pillow*? See what is out there. A few tools we like are Google Keyword Planner (free) and WordStream's Keyword Tool (free). When you type in your word or words, they will also give you some related suggestions to consider.

SpyFu (SpyFu.com): This keyword tool checks out your competitors and gives you a peek at what keywords they are using.

WordStream (WordStream.com): There are a lot of keyword tools out there to help you narrow down the terms you'd want to bid on in Google Ads, but we like WordStream. There is a free version and a paid upgraded version. They have a lot of free resources on their website that are worth a read.

Legal Resources

American Bar Association (AmericanBar.org): You should expect to pay for good legal help, especially for important things like filing patents and other intellectual property issues. However, as you start your business you might need legal help for many issues large and small: signing a lease, getting a sample nondisclosure agreement for contractors, signing partnership agreements. The ABA can be a good place to start when looking for a lawyer—and you can also look to see what free resources are available in your state. There are services for low-income individuals, but sometimes there are "Ask a Lawyer" sites or local events. You might find someone willing to help with simple matters.

LegalZoom (LegalZoom.com): Disclosure: We are not giving out legal advice! LegalZoom has set pricing at the outset of your project and offers self-guided (i.e., DIY) and attorney-guided services, ranging from setting up your company to protecting intellectual property.

Rocket Lawyer (RocketLawyer.com): Disclosure again: We are not giving out legal advice! Laws differ from state to state, and your best bet is always to have local counsel. Rocket Lawyer is an online service that connects you with a local attorney who can do a bunch of things quickly and efficiently, such as setting up a corporation, partnership, or other business entity in your state. For simple things like having a lease reviewed, it can be fast and cost effective.

Patents and Intellectual Property

US Patent and Trademark Office (uspto.gov): You'll want to protect your idea. Getting a patent (or trademark) is very important. The US Patent and Trademark Office has a ton of useful information on its website, and you can conduct a search to find patents for other products that might be similar to yours.

Human Resources

Employee versus contractor: Learn more than you ever wanted to know about whether your newest hire should be an employee or a contractor at the Internal Revenue Service site, irs.gov.

Salary research: When you are making that job offer, do some benchmarking using websites that aggregate salary data by position and location, such as Salary.com or Glassdoor.com.

Manufacturing Help

Alibaba (Alibaba.com): This is a global trade platform and can be a good starting point to find manufacturers who make products like yours, mostly in China. There is a Contact Supplier button so you can reach out to manufacturers directly and start a conversation about your needs.

Thomas (ThomasNet.com): This is essentially a directory of a half-million US-based suppliers of various materials and products. The search function is excellent, and you can search by type, geography, and even type of ownership.

E-Commerce Platforms

Magento (Magento.com): This platform is not nearly as plug-and-play as Shopify and WooCommerce, but it does allow you to more elegantly incorporate nifty features such as selling customized products, selling B2B (wholesale), and managing multiple stores. In addition, it is infinitely customizable (assuming you are willing to invest the time and resources) and also has integrations that can help manage your supply chain.

Shopify (Shopify.com): This out-of-the-box e-commerce platform has a dizzying array of templates to pick from (some are free, and some are third-party generated and have a flat fee), and an equally dizzying selection of plug-in apps and integration to make your online store even smarter and more efficient. The site includes a directory of Shopify developers who can give you bids on setting up your website. Why do you need a developer when it is plug-and-play? Shopify templates use Shopify's own programming language called Liquid, which means that customization of any type requires some external help.

Squarespace (Squarespace.com): This company was initially developed for content-based website creation but has since incorporated the ability to sell products online. Squarespace's claim to fame is the ability to create beautifully designed websites, but as it hasn't been as widely adopted as other platforms for e-commerce purposes, it only offers a few apps and integrations, which may affect your ability to create an efficient system that can grow with you.

WooCommerce (WooCommerce.com): This plug-in works with WordPress, and the themes (designs) are easy to use and customize for anyone who knows these generic coding languages: HTML, CSS, PHP, and JavaScript (this isn't us, but it could be you!). Unlike Shopify, WooCommerce does not include hosting services and subdomains. However, also unlike Shopify, you have complete control over your data and how it is presented.

Crowdfunding Platforms

Note: These are only a few of the *many* platforms out there. We've included the two largest plus one that is exclusively geared toward women.

iFundWomen (iFundWomen.com): This is a crowdfunding platform strictly for women-led start-ups. Unlike other platforms, iFund-Women offers a more holistic approach to bringing a product to market with "expert coaching, professional creative production, a collaborative entrepreneur community, and access to industry connections critical to launching and growing businesses."

Indiegogo (Indiegogo.com): Compared to Kickstarter, Indiegogo distinguishes itself by automatically approving all campaigns for listing. Another feature is the possibility of "flexible funding," which means that even if you do not reach your goal amount, you can still collect what you raised, while Kickstarter is all or nothing.

Kickstarter (Kickstarter.com): The best-known crowdfunding platform, Kickstarter attracts not only highly innovative campaigns, but also a community of committed backers who are willing to take a small risk to be the first to have a new product. Kickstarter vets all campaigns and requires a working prototype, so the products that launch on it tend to be more developed than on other platforms.

Testing and Inspection

When it is time to get your product professionally tested, two of the biggest names are Bureau Veritas and SGS. Both companies are globally known and will test everything under the sun for you.

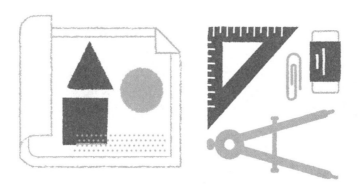

The InventHers and Their Inventions

Sarah Blankinship
Siva Patch (SivaPatch.com)

Chez Brungraber
Gobi Gear (GobiGear.com)

Amy Buckalter
Pulse (LoveMyPulse.com)

Fran Dunaway
TomboyX (TomboyX.com)

Judy Edwards
Squatty Potty (SquattyPotty.com)

Lisa Fetterman
Nomiku (Nomiku.com)

Gloria Hwang
Thousand (ExploreThousand.com)

Stephanie Lynn
Sweet Spot Skirts (SweetSpotSkirts.com)

Michele Mehl
Excy (Excy.com)

Leslie Pierson
GoodHangups (GoodHangups.com)

Sirena Rolfe
Tempus Hood (TempusHood.com)

Nancie Weston
Grayl (Grayl.com)

Mina Yoo
Heroclip (MyHeroclip.com)

Worksheet: Your Conversation with the First Five

Photocopy this worksheet before any conversation with your First Five. For more on these conversations, see Preparing for Your Conversation with the First Five on page 16.

HAVE YOU SEEN A PRODUCT LIKE THIS BEFORE?

_____

_____

_____

_____

DO YOU THINK IT WOULD BE USEFUL FOR A _____

WHO YOU THINK WILL
USE THE PRODUCT

WHY OR WHY NOT?

_____

_____

_____

_____

WOULD YOU PAY FOR IT?

_____

WHAT WOULD YOU PAY FOR IT?

_____

_____

GENERAL FEEDBACK:

_____

_____

_____

_____

MISCELLANEOUS NOTES:

_____

_____

_____

_____

Worksheet: Manufacturer Comparison

When you are ready to start comparing manufacturer facilities, photocopy this comparison worksheet to assess and evaluate their different strengths and weaknesses. For more on this, see Things to Learn About Your Manufacturer (page 129).

MANUFACTURER 1 _____

MANUFACTURER 2 _____

MANUFACTURER 3 _____

MINIMUM ORDER QUANTITY

1 _____

2 _____

3 _____

UNIT PRICE

1 _____

2 _____

3 _____

RECOMMENDED
PRODUCTION METHODS

1 _____

2 _____

3 _____

BOM PROVIDED?

1 _____

2 _____

3 _____

PAYMENT TERMS

1 _____

2 _____

3 _____

MATERIALS USED

1 _____

2 _____

3 _____

TOOLING COST

1 _____

2 _____

3 _____

TOOL CAPACITY

1 _____

2 _____

3 _____

TIME TOOLING WILL TAKE

1 _____

2 _____

3 _____

LIFE SPAN OF TOOL

1 _____

2 _____

3 _____

**NUMBER OF MACHINE X
(MACHINE USED FOR YOUR
PRODUCT) OWNED**

1 _____

2 _____

3 _____

**NUMBER OF OPERATORS
FOR MACHINE X**

1 _____

2 _____

3 _____

**NUMBER OF UNITS
THEIR FACILITY CAN
PRODUCE PER DAY/WEEK**

1 _____

2 _____

3 _____

OTHER CUSTOMERS

1 _____

2 _____

3 _____

PARTS TO BE MADE IN-HOUSE

1 _____

2 _____

3 _____

PARTS TO BE OUTSOURCED

1 _____

2 _____

3 _____

Worksheet: Profit and Loss Statement

Income

Revenue

Discounts _____

Net Revenue _____

Cost of Goods Sold (COGS)

COGS

Total COGS _____

Gross Margin _____

Operating Expenses

Research & Development (R&D)

Product Design & Development _____

Prototyping & Testing _____

Tooling _____

IP Protection _____

Total R&D Expense _____

Sales & Marketing

Content Creation _____

Advertising _____

PR _____

Social Media _____

Trade Shows _____

Customer Service _____

Total Sales & Marketing _____

General & Administration (G&A)

Payroll & Benefits _____

Software & Equipment _____

Professional Services (Legal, HR) _____

Insurance _____

Meals & Entertainment _____

Office Supplies _____

Bank Fees _____

Occupancy (Office Rent, Utilities) _____

Travel _____

Total G&A _____

Total Operating Expenses _____

Earnings Before Interest, Taxes, Depreciation, and Amortization (EBITDA) _____

EBITDA Margin _____

Index

About the Authors

MINA YOO is a savvy, seasoned entrepreneur and currently the CEO of Heroclip, a consumer brand that makes products to simplify and organize daily life. In her previous life, she was a professor and researcher of entrepreneurship at the University of Washington Foster School of Business and Stanford University. Mina, her company, and her products have been featured in GeekWire, ESPN, *Boston Globe*, *Wall Street Journal*, *Health*, *Men's Health*, NBC affiliate King 5, GearJunkie, and *Outside* magazine, among many others. Mina has a dual PhD in business and sociology from the University of Michigan, and a BA from Brown University. Mina's favorite activities include CrossFit; traveling; playing with her kids, Kai and Mila; wine tasting with her husband, Mark (without her kids!); and coming up with new ideas for products that help families hang out together.

HILARY MEYERSON is a professional writer and marketing consultant, and holds a BA from Middlebury College and a JD from the University of Washington. She is a regular contributor to *Seattle* magazine, and has been a writer in residence at Hedgebrook, Soapstone, and Caldera Arts. Her work has also appeared in *Seattle Met*, *Seattle Business* magazine, and *McSweeney's*, among others. She also is the founder of a social marketing company, Little Candle Media, where she helps start-up businesses make the most of their marketing efforts. In her free time, she is usually found doing long urban rambles with her husband, Randy; watching bad Netflix romcoms with her daughter, Harper; geeking out over rocket news with her son, Henry; or solo hiking in the Central Cascades.